AF345764

Long-term Management of Nuclear Fuel Waste: Implications of Plutonium and Americium Radiotoxicity

Meenu

Copyright © 2024 by Meenu

All rights reserved. No part of this book may be reproduced in any manner whatsoever without written permission except in the case of brief quotations embodied in critical articles and reviews.
First Printing, 2024

Content

1. Motivation

The disposal of waste produced by the commercial usage of nuclear fuel is a very demanding task due to the high radiotoxicity of the SNF. Transuranium elements, such as Pu and Am are formed in neutron capture reactions during reactor operation. The neutron capture of ^{238}U, which is present in a large abundancy in nuclear fuel, followed by two beta decays produces the transuranium isotope ^{239}Pu. Through consecutive neutron capture reactions of ^{239}Pu, followed by a beta-decay, the radionuclides ^{241}Am (2 neutrons captured, followed by beta decay), and ^{243}Am (4 neutrons captured, followed by beta-decay) are formed. When the fuel rods are removed from the reactor core, this SNF will have accumulated approximately 1% Pu and a very modest amount of Am. However, over time the amount of Am in SNF will increase, as the decay of ^{241}Pu through beta emission will produce the ^{241}Am isotope. Therefore, both Am, and Pu will govern the long-term radiotoxicity of SNF from about 500 to 1 million years.

The SNF will either be directly disposed of or reprocessed where uranium and plutonium are recovered from the waste and reused as mixed oxide (MOX) fuel. The remaining waste will nevertheless still require disposal over extended periods of time.

For the long-term disposal of SNF or HLW streams from reprocessing operations, deep-geological disposal has been considered the most promising concept. The deep subsurface environment should isolate the problematic waste nuclides until they have decayed. While the ingress of water to such an underground disposal site cannot be excluded over time scales of hundreds of thousands of years, a multi-barrier concept comprising both natural and engineered barriers has to be designed to prevent migration of radionuclides, such as ^{239}Pu, ^{241}Am, and ^{243}Am, into the environment of the repository site. The multi-barrier concept comprises several solid phases, which show good properties towards the removal of actinides from the aqueous phase via sorption and/or incorporation reactions on and into the solid. The extent of removal of the radioactive contaminants is not just determined by the presence or absence of these solid phases but is connected to the speciation of the actinides under the prevailing conditions, which largely depends on their oxidation state. In reducing subsurface environments, the long-lived actinide elements are present in their tetravalent and trivalent oxidations states, of which the latter one is important both for Pu and Am. Further, the trivalent oxidation state is more soluble than the tetravalent one and can therefore be considered more mobile. For this reason, the focus of this book has been on trivalent actinides and their sorption and incorporation reactions in the subsurface environment.

As solid phase for these interaction studies, zirconium oxide or zirconia (ZrO_2) has been focused upon. It is present as a corrosion layer on the Zircaloy cladding material surrounding the SNF rods and will therefore be one of the first interaction partners and a potential immobilization barrier for mobilized Pu and Am from the SNF waste matrix. Further, zirconia may be an important crystalline waste form for disposal concepts for HLW streams from reprocessing activities. Alternatively, it could be considered for safekeeping of dismantled plutonium to reduce proliferation and criticality risks associated with the ^{239}Pu isotope, in particular.

Despite the role of zirconia in the various disposal strategies, very few studies can be found on the sorption of actinides on zirconia[1–5] especially when focusing on the trivalent oxidation state for which no publications are available. Incorporation studies are also scarce and mostly conducted from a radioactive waste conditioning point of view, i.e. where actinide immobilization within the zirconia ceramic is accomplished in high-temperature calcination or sintering procedures aiming at the long-term storage of radionuclides within the ceramic matrix.[6–10] Incorporation processes occurring in a waste repository as a result of dissolution and recrystallization reactions have not been taken into account in such studies.

Therefore, this book focuses on closing the knowledge gaps on the sorption of trivalent actinides on zirconia, by studying the macroscopic retention as well as the speciation of trivalent actinides, using Eu^{3+} and Cm^{3+} as representative for Pu^{3+} and Am^{3+}, on zirconia. The incorporation investigations in this book build on the aforementioned existing studies, however, a systematic approach was pursued in a large range of trivalent doping from 25 ppm (0.0025 mol%) to 26 mol% applying a combination of bulk structural methods, such as PXRD, SEM, and TEM, as well as techniques to study the local structure in these phases, such as TRLFS and EXAFS. Furthermore, this book extends the scope of research to conditions present in a HLW repository, where untreated burned up fuel rods are stored and where dissolution, co-precipitation, and subsequent actinide incorporation reactions in zirconia can occur in solution under slightly elevated temperatures.

2. Introduction and Theory

2.1. Historic Background

The Rutherford model of atoms, which was derived by Ernest Rutherford based on the famous Geiger-Marsden experiment, better known as the gold foil experiment, has set the ground for the fundamental structure of atoms as we know it today. More specifically, Rutherford discovered that an atom has a positively charged core, making up only one ten thousandth of the diameter of an atom but almost all of its mass, with a shell of electrons around it where most of the space is empty. However, he was also the creator of the first man-made nuclear reaction through bombardment of nitrogen with alpha-particles at the University of Manchester in 1919. 13 years later, in 1932, James Chadwick, brought to the Cavendish Laboratory of the University of Cambridge by its director at that time, Ernest Rutherford, observed the emission of neutral particles when irradiating beryllium with alpha particles, which they called neutrons. Only 7 years later, Otto Hahn and Fritz Strassmann had observed the induced fission of uranium under neutron irradiation, setting the foundation for the usage of the large amount of released energy in this fission reaction ($\sim$ 200 MeV per fission of one ^{235}U atom) for civil, but also military purposes.[11,12]

In 1954, the Obninsk Nuclear Power Plant in the former Soviet Union was the first grid connected nuclear fission reactor and many have followed since then. The highest number of operating nuclear reactors reached 444 in 2002. In 2011, over 70% of the generated nuclear energy was produced by the "big six" countries, which were, in order of power generated, the United States of America, France, Russia, Japan, South Korea, and Germany.[13] The disaster at the Fukushima Daiichi nuclear power plant in Japan on 11th of March 2011 has led to drastic changes in the energy production for many countries. The number of operational nuclear reactors in Japan has dropped from 54 in 2010 to 5 in 2017. On the 15th of March, only 4 days after the Fukushima accident, plans for the extension of German reactors was abruptly put on hold and a review of the German energy strategy was initiated. As a result, the immediate shutdown of eight power plants was decided and phase out of the remaining 9 operating reactors was scheduled until the end of 2022. The decision of shutting down all German power plants has enforced the national search for a final repository for HLW. This search was originally started in 1999, where, for the first time, scientifically sound criteria for the search of a final repository for HLW were gathered.[14] Until today, 20 years later, a decision about a repository location has not been made.

However, it was decided that the host formation should be rock salt, clay, or crystalline rock and that a decision on a repository location for HLW should be sought until 2031.[15,16] To find the most suitable host-rock and to be able to evaluate the performance of a repository for HLW built in the given host-rock formation, fundamental understanding of geochemical processes such as the interaction of released RNs from the SNF matrix with the deep disposal environment as well as reliable thermodynamic data of these retention mechanisms will be needed.

2.2. Disposal of Spent Nuclear Fuel (SNF)

2.2.1. Composition of SNF

SNF is nuclear fuel, which has been removed from a nuclear reactor core after being irradiated for a certain time. While pure UO_2-based nuclear fuel typically consists of 97% ^{238}U and about 3% of fissile ^{235}U in case of a light-water reactor, the neutron irradiation produces many additional nuclides through fission, as well as neutron capture reactions. SNF still consists of 95% ^{238}U, less than 1% ^{235}U, ~ 3% fission products, ~ 1% plutonium, ~ 0.5% ^{236}U and 0.06% of the minor actinides (MA) neptunium (Np), americium (Am), and curium (Cm) (Figure 1). The activity of SNF is extraordinarily high when freshly taken out of a nuclear reactor. The exact activity depends on the reactor type as well as the burn-up of the fuel, however, a typical value of the activity of a metric ton of SNF would be 10^{17} to 10^{18} Bq.[17]

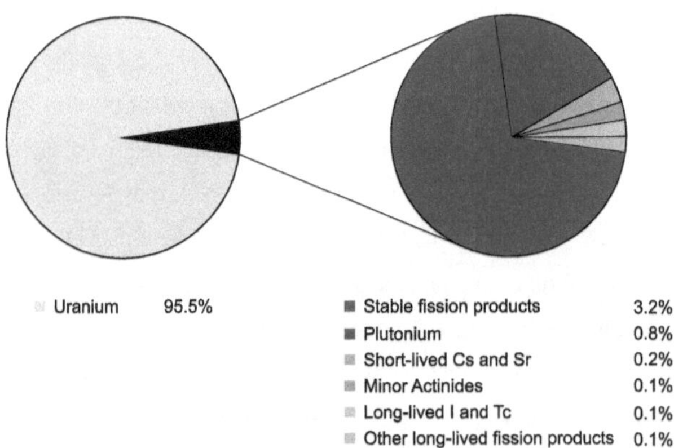

Figure 1: Composition of SNF from a standard pressurized water reactor after 10-year cooling. Figure adapted from a Nuclear Energy Agency status report[18].

A large share of this activity originates from short-lived fission products, which is why the activity is about 100 times lower after a few years of storage in the cooling ponds. However, it takes more than 100 000 years until the activity of SNF is close to the level of the natural uranium used as nuclear fuel. Due to this long-term radiotoxicity of SNF, differing waste treatment pathways have been considered, such as the reprocessing of SNF or the direct disposal of the whole fuel rod assemblies. In the latter case the SNF assemblies from the storage pools are put into dry cask storage, as most of the heat generating isotopes have decayed by then.

From about 500 years to 1 million years after the SNF has been removed from a reactor core, plutonium and its decay products as well as the minor actinides and their decay products dominate the remaining activity (Figure 2). The term *minor actinides* is commonly used in the context of SNF for all actinides present in SNF besides uranium and plutonium, which are called the *major actinides* due to their higher abundance in SNF. Especially ^{239}Pu ($t_{1/2}$ = 2.41 · 10^4 a), ^{243}Am ($t_{1/2}$ = 7370 a), and ^{241}Am ($t_{1/2}$ = 432.2 a) with their rather long half-lives, dominate the long-term radiotoxicity, despite of their rather small fraction in SNF.

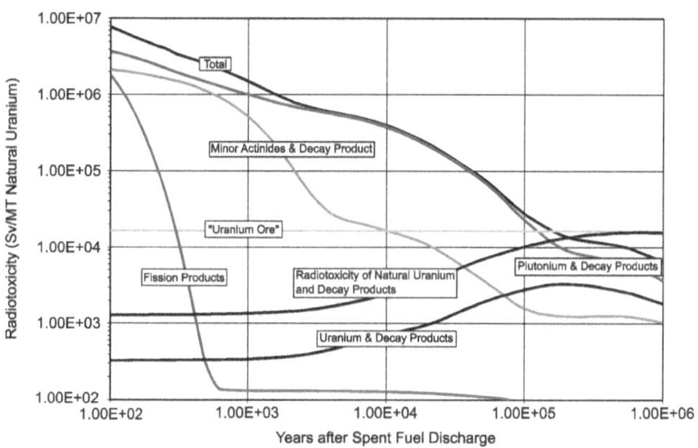

Figure 2: Radiotoxicity of SNF in dependence of time after fuel discharge. Figure adapted from a Nuclear Energy Agency status report[18].

2.2.2. Disposal Strategies for SNF

2.2.2.1. Direct Disposal of SNF

Even though many disposal concepts for SNF have been considered, e.g. sea disposal or disposal in outer space, the only commonly accepted way according to the *World Nuclear Association*, is deep geological disposal in a host rock. This has many advantages, one being that once such a repository is built and all HLW is stored in it, it can be sealed without the need of actively maintaining the facility afterwards. Furthermore, the deep geological disposal provides the opportunity of isolating the SNF through multiple release barriers, which will prevent the discharge of radionuclides into the surrounding environment.[19]

This multi-barrier concept is thought to consist of a technical barrier, i.e. the containment of the radioactive waste, a geotechnical barrier, which is the backfill used to seal the repository and the geological barrier, i.e. the host rock formation itself. An exemplary concept is presented in Figure 3, illustrating the Swedish final waste repository, designed and operated by the Swedish nuclear fuel and waste management company SKB. Here, the SNF within the cladding material is positioned in a copper canister (technical barrier) which is placed in a borehole in a crystalline bedrock (geological barrier) and filled with bentonite clay (geotechnical barrier).

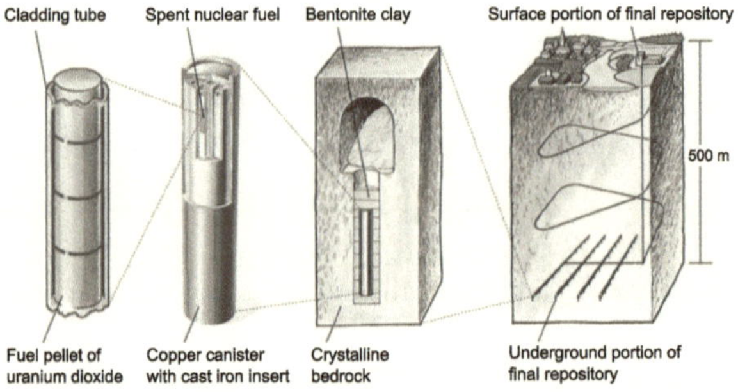

Figure 3: Multi-barrier concept design, for the Swedish bedrock repository designed and operated by SKB.[20]

Such a multi-barrier concept is expected to minimize any possible disruption to the storage facility in the far future and thereby ensure safe storage of HLW for a long time. Such disruptions include the ingress of water which is considered a worst-case scenario in any deep-geological HLW repository as it can lead to oxidation processes. Oxidative dissolution of the spent nuclear fuel matrix could release radionuclides to the surrounding where they could be further mobilized with the flowing ground water. Therefore, the direct disposal of SNF must be accompanied by thorough safety assessments comprising both the natural and engineered barriers as well as natural phenomena such as earthquakes and ice age cycles. Nature itself gives examples where such geological isolation has worked over long periods. A very striking example is Oklo in West Africa, where 2 billion years ago, several natural spontaneous nuclear reactors operated within a uranium ore vein. These natural nuclear reactors could operate for about 500 000 years, due to the sufficient natural abundance (~ 3%) of the fissile U-235 isotope in the natural uranium ore present at that time. In this period, the natural reactors produced a large amount of highly active radionuclides, more specifically 5 tons of fission products and 1.5 tons of plutonium, which remained at the site until they finally decayed.

An additional challenge to this disposal pathway is the retrievability and recoverability of SNF at a later stage in time, which will be compulsory, for example in Germany, for a minimum time-frame of 500 years after sealing of a final waste repository.[21]

2.2.2.2. Reprocessing of SNF

Most of the SNF consists of Uranium, i.e. ^{238}U, which has a very low specific activity and radiotoxicity and a small fraction of fissile ^{235}U, which was not consumed during reactor operation and which can be re-used for fuel production. Further, many specific long-term safety problems originate from plutonium and the minor actinides within the uranium matrix in SNF. Therefore, re-processing of the SNF with the aim of separating the re-usable ^{235}U and the low-toxic ^{238}U from the more problematic fractions has some advantages with respect to the disposal of SNF. The *PUREX* (Plutonium-Uranium Refining by Extraction) process, which is based on a liquid-liquid extraction method, can be employed for the separation of the actinides from SNF. So called mixed oxide (MOX) fuels are produced from the recovered uranium and plutonium while the remaining minor actinides and the fission products are solidified in other waste matrices, such as borosilicate glass[22] or cement[23]. As ^{238}U is removed in the reprocessing, the overall HLW volume is significantly decreased in the process. The long-term radiotoxicity of the solidified waste is reduced, but due to the presence of trace concentrations of plutonium, several long-lived fission products, and the minor actinides in the waste effluent, storage of the waste forms for long time spans is still required. Therefore, to further reduce the concentration of radionuclides in the glass/cement matrices, various partitioning and transmutation strategies have been considered.

Due to the large effort, high costs and risks in treating the highly active SNF in addition to the production of weapons grade plutonium in the reprocessing process, this a highly debated topic. Even more so, as direct disposal of radioactive waste streams into the surrounding environment has been done in some early reprocessing facilities. In the reprocessing facilities in Sellafield (UK) for example the mean monthly discharge of radioactive effluents into the Irish sea in 1959 and 1960 was 72.5 TBq.[24] Furthermore, the reduction of the volume of HLW comes with the cost of largely increasing the volume of low and intermediate level waste.[25] To date the countries China, France, India, Japan, Pakistan, Russia, and UK are operating reprocessing facilites.[26] However, other countries, such as Germany, have used the reprocessing facilities of France and the UK until 2005 for reprocessing of SNF and, therefore, have reprocessed SNF to consider for storage in a HLW repository.[27]

While many countries have not decided upon a waste management strategy yet, Germany as well as Sweden and Finland now favor a disposal of the complete fuel assembly without reprocessing.[28–30]

2.2.3. Actinide Speciation in a HLW Repository

As described before ^{239}Pu, ^{241}Am, and ^{243}Am are a major concern for the long-term safety of a HLW repository. Their mobility strongly depends on various factors such as their oxidation state, their speciation as well as their chemical state, i.e. whether a solubilized species or a solid phase is present.

The oxidation state of the actinides depends on the ambient conditions present in the repository. A fully sealed repository is expected to exhibit reducing conditions, which is advantageous as it hinders corrosion processes of the technical barriers, i.e. the canisters, or other constituents of the multi-barrier concept. Water which might access such a repository at some point in time will, however, bring potential for oxidation with it due to oxygen as well as salts solubilized in the water. How much the reducing environment in the repository is affected by such a perturbation of the barrier depends on the amount of water, as well as the origin of the water and the amount of oxygen accessing the repository through, for example cracks in the host material.

Americium in solid phases is stable as Am^{4+}, while americium complexes in aqueous solutions prevails the oxidation state 3+. Plutonium is a very redox sensitive element which can occur in various oxidation states. At reducing conditions, however, Pu^{4+} and especially Pu^{3+} are likely to form. As a result, the mobility of trivalent actinides in the surrounding of a repository governs the spreading into the environment. Therefore, the trivalent actinides and their lanthanide analogues have been the focus in this book.

Due to the activity and costs generated by experiments performed with trivalent actinides like Am^{3+} or Pu^{3+} and due to the challenging redox chemistry of especially the latter one, analogues are favorable to study their retention behavior. For this, Cm^{3+} can be used since it has a comparable ionic radius of 97 pm (at coordination number (CN) 6) as compared to 97.5 pm and 100 pm (at CN 6) for Am^{3+} and Pu^{3+}, respectively. Cm^{3+} shows exceptional luminescence properties which can yield crucial information on its coordination environment, as will be discussed in detail in chapter 2.4.2.1. Further, it is redox insensitive, and the nuclide used in this book, i.e. ^{248}Cm, has a low specific activity resulting from the long half-life ($t_{1/2}$ = 348 000 a). Especially this isotope is, however, extremely rare and, therefore, extremely expensive which is why lanthanide analogues and yttrium were used for most experiments performed in this book and only selected samples were prepared using

Cm^{3+}.

Natural europium (i.e. 48% ^{151}Eu, 52% ^{153}Eu) is a very useful analogue, due to its stable oxidation state of 3+ in aqueous conditions as well as its excellent luminescence properties, as will be described in more detail in chapter 2.4.2.1. Its ionic radius is 95 pm at CN 6 and, thus, close to the ones of Am^{3+} and Pu^{3+}.[31] Furthermore, it is inexpensive as well as inactive and, therefore, ideal for studies where a large concentration of trivalent cations is needed, such as incorporation studies.

In some experiments trivalent yttrium was used as a substitute for Eu^{3+} due to a more advantageous X-ray absorption edge energy as well as to avoid energy transfer effects, hindering the analysis of luminescence processes. Its only stable oxidation state is 3+, its ionic radius is 90 pm at CN 6 and it is inactive and inexpensive, making it a good analogue for Eu^{3+} and the trivalent actinides.

2.2.4. The Role of Zirconia in SNF Disposal Concepts

In the following chapters, zirconium oxide or zirconia (ZrO_2) will be discussed in detail as it plays a central role both in the context of direct disposal of SNF as well as in various stages of SNF reprocessing, as a potential immobilization matrix for specific waste streams and as a target material in transmutation.

As previously mentioned, reprocessing of SNF and the generated HLW opens possibilities for various partitioning and transmutation strategies. Such separated waste streams containing e.g. the minor actinides or plutonium from dismantled nuclear weapons could be immobilized in specific, tailored materials such as ceramics. While many approaches have been undertaken to create synthetic rock materials, called SYNROC[7], also natural minerals are being considered for the conditioning of radioactive waste, like monazite,[32] zirconolite,[33] perovskite,[34] titanite,[35] or zirconia.[6] Zirconia has received attention as a ceramic host for the incorporation of actinides due to its very low solubility and high radiation tolerance.[36–40] In addition, zirconia is capable of incorporating large quantities of both trivalent and tetravalent actinide ions into the host-lattice, which are the most important oxidation states of the radionuclides present in the above-mentioned HLW streams. Furthermore, zirconia is studied in the research field of inert matrix fuels (IMF), where it was originally intended for the improvement of fuel properties.[41] More recent studies investigate the usage of zirconia based IMF for the transmutation of excess plutonium from weapons dismantling and SNF reprocessing, or its usage as target material for the transmutation of minor actinides.[42,43]

Also, for repository concepts with direct disposal of SNF, zirconia is of high relevance. ZrO_2 is the primary corrosion product of Zircaloy cladding material surrounding nuclear fuel rods, and is, therefore, the first interaction partner for mobilized RNs. The cladding material has been designed to withstand both high irradiation doses as well as corrosion when in contact with water, which is already of importance during reactor operation and during interim storage of SNF in storage pools. Despite of very good corrosion resistance, Zircaloy corrosion will eventually occur in aqueous solution, resulting in the formation of a ZrO_2 corrosion layer, deposited on the intact Zircaloy cladding (Figure 4).[44,45] Additionally, cracks can occur in the corroded cladding material facilitating the mobilization of actinides from the uranium matrix.

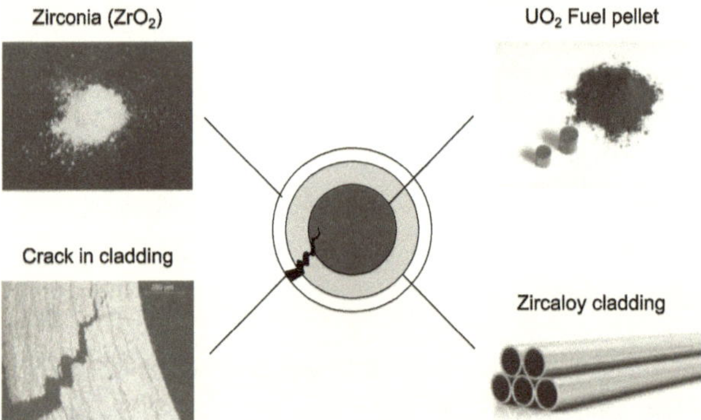

Figure 4: Scheme of a SNF rod and images of a UO$_2$ fuel pellet[46] in the center (black), the Zircaloy cladding[47] around it (grey), and a corrosion layer of zirconia on top of it (white). A crack in the cladding material[48] is indicated with a black shape on the bottom left.

In a final HLW repository where water has come into contact with the UO_2 ceramic and surrounding fuel rods, dissolution of Zircaloy and the formation of zirconia on the cladding surface may be accompanied by recrystallization reactions with dissolved radionuclides from the SNF matrix. Such processes could take place for example in a crack of the Zircaloy cladding material. These reactions may involve incorporation of actinides within the zirconia corrosion layer or immobilization of radionuclides through sorption onto the oxide surface. Thus, from a nuclear repository safety aspect, the interaction mechanisms between especially the actinide elements and zirconia should be explored.

2.3. Interactions of Actinides with Zirconia

2.3.1. Actinide Sorption on Zirconia

When dissolved trivalent actinides come into contact with their surroundings, such as the cladding material or any of the barriers around the HLW itself, actinides can adsorb to their surfaces via two processes, outer-sphere sorption and inner-sphere sorption, where sphere stands for the hydration sphere around the cation. When actinides are loosely bound or sorbed to a surface solely through *Coulomb* interactions, the hydration sphere of the cation remains pristine and outer-sphere sorption, also called physisorption takes place. When chemical bonds form between the surface and the RN, the hydration sphere is partly removed and inner-sphere sorption, also called chemisorption takes place.

Mineral surfaces in natural environments typically exhibit a hydrated surface layer. In case of zirconia various types of hydroxyl groups exist at the surface, i.e. primary, secondary and tertiary groups (Figure 5) differing in their accessibility as well as their pH-dependent protonation/ de-protonation behavior.[49] With increasing pH the surface charge decreases due to the deprotonation of the surface hydroxyls. At a certain point, the overall surface charge is zero since the amount of positively charged surface groups equals the amount of negatively charged ones. Various values can be found in literature for this so called isoelectric point (IEP or pH$_{IEP}$), ranging from 4 to 10, resulting from the rather complex behavior of the zirconia surface.[50]

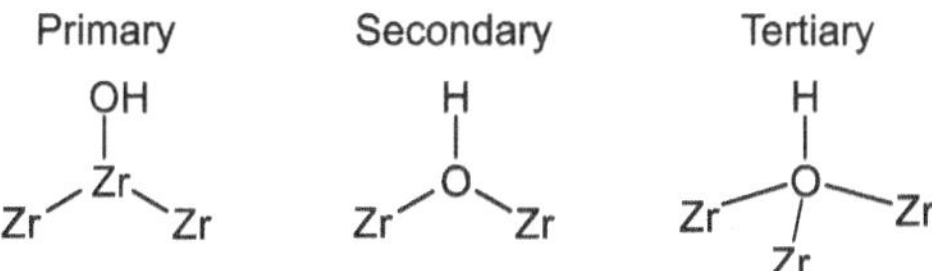

Figure 5: Primary (left), secondary (middle) and tertiary (right) surface hydroxyl groups on a zirconia surface.

With the deprotonation of the zirconia surface at increasing pH, the positively charged actinide cations are more and more attracted by the surface of zirconia, leading to a steep increase of sorption and, therefore retention of actinides from a certain pH onwards. To quantify the pH-dependent adsorption reactions, so called batch-sorption experiments were performed, which monitor the removal of the actinide from solution in the presence of zirconia as a function of pH.

The key factor for the retention of actinides is the durability of a retention process towards changes in the system. An outer-sphere complex of an actinide can already be removed by a change of the ionic strength in the aqueous phase. This is due to an increase of the concentration of other cations, such as Na^+, K^+ or Ca^{2+}, competing with the actinide for the sorption site. For the re-mobilization of an inner-sphere complex, a more drastic change, such as a decrease of the solution pH needs to take place. Furthermore, the inner-sphere sorption can form the foundation for a later incorporation into a mineral. While batch-sorption experiments can give valuable quantitative information on the sorption behavior on a mineral surface, limited molecular information on the type of surface interaction can be gained from them. Therefore, spectroscopic methods are needed to gain deeper insight into the sorption mechanisms in a given system.

Many studies can be found for the interaction of actinides with various clay minerals in host rock formations[51–58] or in envisioned buffer and back-fill materials[59–74]. In addition several studies report on actinide sorption reactions with crystalline (host) rock[75–78], various iron minerals as corrosion products from SNF steel canisters[79–83] and aluminum-bearing solids as representatives for aluminosilicates[84–88] or iron containing solid phases[79,89–92]. However, the interaction of radionuclides, especially the actinides, with the zirconia surface in aqueous systems has received very little attention. Some studies have been performed on the sorption of U(VI) on zirconia[1–3] while only two studies exist on the sorption of transuranium elements, i.e. Pu(IV)[4,5] and Np(V)[5]. However, no studies can be found on the sorption of trivalent actinides on zirconia, despite of their high relevance for safety considerations of a HLW repository.

2.3.2. Actinide Incorporation into Zirconia

Incorporation of actinides into a solid can take place when materials crystalize in the presence of an actinide or via the equilibrium dissolution and recrystallization processes at a surface. An incorporated cation can only be re-mobilized through dissolution of the retaining material, which is why this is the strongest form of retention. Incorporation is also the sole mechanism of actinide association in various waste forms, which is why detailed understanding of the actinide solid solution process in these solid phases is mandatory.

The most thermodynamically stable phase of zirconia is the monoclinic crystal structure with the space group $P2_1/c$, which transforms into the tetragonal ($P4_2/nmc$) and the cubic phase ($Fm\bar{3}m$) at high temperatures of around 1200 °C and 2370 °C, respectively. Whilst the cubic phase is isostructural to the fluorite type structure, the tetragonal and monoclinic phases show distorted structures from that.[93,94] It should be noted that this is only true for crystallites larger than about 30 nm. For particles smaller than 30 nm, the tetragonal structure occurs and for particles smaller than around 15 nm the cubic structure has been seen to be the stable phase.[95-97] The high temperature phases, tetragonal and cubic zirconia, can be stabilized at room temperature by the formation of solid solutions with foreign ions. To understand this stabilization mechanistically one needs to know the reason why the high temperature phases are instable at ambient conditions. In the monoclinic system zirconium has an oxygen coordination number of seven, while in tetragonal or cubic symmetry eightfold coordination exists. Regarding that zirconium(IV) has a comparably small ionic radius (78 pm in sevenfold coordination) it can be understood that an eightfold coordination of oxygen atoms is electrostatically unfavored. Two different stabilization mechanisms have been described in literature.[98-107]

(i) The addition of under- or oversized cations into the monoclinic crystal lattice leads to a distortion of the lattice, so that the oxygen atoms order themselves in two coordination shells of four oxygens each with distinct bond lengths which leads to a higher oxygen-oxygen distance, subsequently reducing their repulsive interactions.

(ii) The second mechanism is the addition of aliovalent dopants such as An^{3+}, which create the need for charge compensation. The charge compensation takes place via the inclusion of oxygen vacancies into the crystal lattice. This oxygen vacancy has been observed to prefer the nearest neighbor (NN) position of the zirconium ion, reducing its coordination number to seven.

Since all actinides are largely oversized as compared to the Zr^{4+} cation, and since especially the trivalent actinides are of importance for studies in the field of HLW repository research, both mechanisms will work synergistically here, to stabilize the high-temperature zirconia polymorphs. Solid solutions of zirconia have found various other applications because of the combination of strength, fracture toughness, low thermal conductivity and high ionic conductivity. Examples are the usage in devices like gas sensors,[108,109] solid oxide fuel cells (SOFCs),[110,111] metal-insulator-semiconductors[112] and several more. Besides the technical and electronical usage, biomedical applications are used frequently, especially as teeth implants. Because of that, the influence of parameters like the temperature and the dopant cation on the mechanical properties has been investigated intensively.[113–117] However, very few publications aim for a mechanistic understanding of the solid solution formation and stabilization process. The lack of systematic studies of the crystal phase as a function of the used stabilizer content is another reason why the behavior of zirconia is up to date not well understood.

2.4. Characterization of Interactions of Trivalent Actinides with Zirconia

2.4.1. Bulk and Surface Characterization Methods

Actinide adsorption reactions on crystalline materials like zirconia, are governed by various properties such as the available surface area or adsorption sites, the surface charge, and the reactivity of the surface toward other dissolved ionic or molecular species present in the investigated system. Various methods exist to address these material properties. In the present work electron microscopic techniques were applied to study the morphology and particle size of zirconia. The N_2-BET method was used to determine the surface area. With zeta-potential measurements the electrophoretic mobility of zirconia particles was derived which allows conclusions about the surface charge. Batch-sorption experiments yield a quantitative measure to the uptake properties of zirconia in dependence of the pH. In this work, powder X-ray diffraction (PXRD) plays a very important role due to its capability of deriving qualitative and quantitative phase composition of single- as well as multi-phase systems. The quantitative analysis is done using *Rietveld* refinement, which is a least-squares refinement using crystallographic data.[118] Furthermore, information on the crystallite size can be derived from the diffraction pattern via the *Scherrer* equation.[119] The *Scherrer* equation makes use of the fact that the full width at half maximum (FWHM) of the diffraction peak of a crystalline material is mainly dependent on the crystallite size. Small crystallites will yield broader peaks than larger crystallites, while an increasing crystallinity of the sample will result in narrower diffraction patterns. Therefore, the *Scherrer* equation (Eq. 1) can be used to calculate the crystallite size from the PXRD pattern.

$$D_{hkl} = \frac{K\lambda}{B_{hkl} \cdot \cos\theta} \qquad (1)$$

D_{hkl} = Crystallite size perpendicular to the lattice planes

hkl = Miller indices of the investigated crystal planes

λ = Incidence wavelength of the used X-ray source

B_{hkl} = FWHM of a certain lattice reflection

θ = Bragg angle

K = Size-dependent Scherrer constant, here 1 was used as approximated value

2.4.2. Spectroscopic Methods

To gain an in-depth understanding of processes taking place on a molecular level, as opposed to processes on a macroscopic scale, various spectroscopic methods can be applied. These address changes occurring in e.g. the nuclear, electronic, vibrational, or rotational states of the actinide or the matter it resides in, following sorption complexation reactions or incorporation into a solid structure.

In the current work spectroscopic techniques focusing on the electronic structure of the actinide and its surrounding have been applied. The main method used here to study the actinide speciation at the zirconia surface and incorporation of Eu^{3+} or Cm^{3+} into the bulk, is TRLFS.

Another method used extensively in this book is extended X-ray absorption fine-structure spectroscopy (EXAFS), yielding element specific information on the average environment of a certain analyte, e.g. Zr^{4+} or Y^{3+}.

These main methods will be described in detail in the following chapters.

2.4.2.1. Luminescence Spectroscopy (TRLFS)

TRLFS is a method to study the electromagnetic emission caused by the transition of electrons from an excited electronic level to the ground state, where a laser is used for excitation. Depending on the multiplicity difference of the excited and the ground state this luminescent transition is commonly divided into fluorescence ($\Delta S = 0$) or phosphorescence ($\Delta S \neq 0$), where spin exchange takes place via intersystem crossing. In these studies, Eu^{3+} and Cm^{3+} have been used as fluorophores, where spin exchange takes place between the emitting transitions. Therefore, phosphorescence would be the correct term which, however, is rather unconventional. The term luminescence will be used in this work from hereon to overcome the inaccuracy of the term fluorescence.

Two general excitation strategies can be applied to induce the luminescence of Eu^{3+}/Cm^{3+}, i.e. UV-excitation (indirect) and direct excitation. When using UV light for the excitation process, an excess amount of energy is provided to the system to reach a strongly absorbing electronic state. Radiationless de-excitation to the emitting state is then followed by luminescent de-excitation to the ground state. In case of direct excitation, a wavelength is used which matches exactly the energy gap between the ground state and the emitting excited state for so called resonant absorption. The advantage here is, that differing species will have slightly differing energy gaps and, therefore, site-selective excitation, where only a specific species is excited, can be achieved. However, this is connected with a larger experimental demand since low temperatures (typically < 20 K) are needed to suppress thermally induced broadening of the energy levels. Furthermore, a tunable laser setup with a good wavelength resolution is necessary. For this, optical parametric oscillator (OPO) laser setups have been commonly used, where an optical non-linear crystal combined with an optical resonator is used to generate the desired output wavelength via a non-linear optical frequency-mixing process. The usage of dye-laser setups has become more and more popular due to a simpler handling and maintenance. For this, an organic dye is excited via a pump laser (here: second harmonic of Nd:YAG) and stimulated emission of the desired wavelength is obtained by using an optical grating.

Eu³⁺ Luminescence Spectroscopy

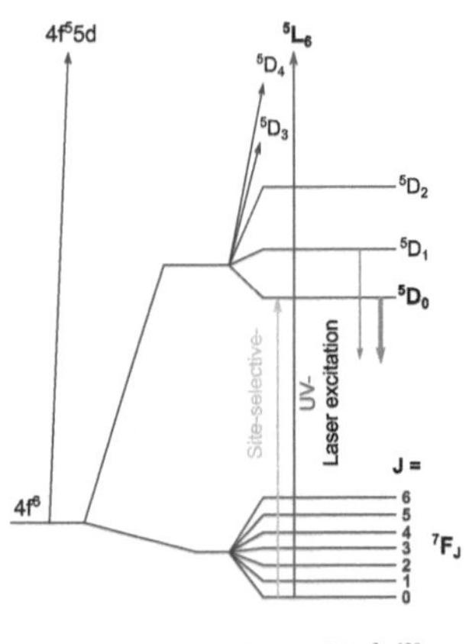

Figure 6: Term scheme of Eu^{3+}.[120]

A schematic term scheme of Eu^{3+} is presented in Figure 6. The excitation of Eu^{3+} is often achieved by inducing a transition to higher energy levels, than the emitting one, since intense absorption peaks can be found in these transitions. The ${}^5L_6 \leftarrow {}^7F_0$ ($\lambda_{ex} = 394$ nm) transition is especially favorable, since it is the most intense transition in the absorption spectrum in the UV region. Radiationless deactivation to the ${}^5D_J$ levels, mainly to the ${}^5D_0$ level, which is the main emitting level, takes place thereafter. The following f-f transitions of ${}^5D_0 \rightarrow {}^7F_J$ are very characteristic and yield manifold information on the Eu^{3+} environment. The f-f transitions are forbidden, in accordance with *Laportes' selection rule*, which is, however, only strict in the gas phase. In a matrix, vibronic coupling as well as mixing of higher configurations leads to a weakening of this rule.[120] An exemplary emission spectrum is presented in Figure 7.

The ^7F_J ground state term is split into 7 levels which can all be observed in emission spectra. Due to several reasons the lowest three ($^7F_{0-2}$) are usually focused on. One reason is the specialty of each of these transitions. The $^5D_0 \rightarrow {}^7F_0$ transition inherits the property of $\Delta J = 0$. Therefore, this transition is strongly forbidden, however, only in a centrosymmetric system. Since the degeneracy of the sub-levels of the ^7F_J levels equals 2J+1, the $^7F_0$ level is non-degenerate and, thus every Eu^{3+} species exhibits only one emission peak. This can help greatly to identify the amount of differing species present in a sample.

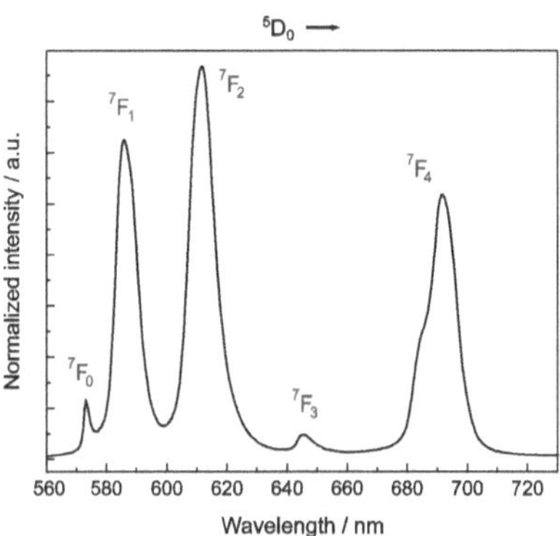

Figure 7: Example emission spectrum of a Eu^{3+} sample in solution after UV-excitation ($\lambda_{ex} = 394$ nm)

The transition to the $^7F_1$ level is the only magnetic dipole transition of the $^5D_0 \rightarrow {}^7F_J$ transitions, and, therefore, independent of the electronic environment. For this reason, this transition can act as an internal standard where Eu^{3+} emission spectra are normalized to its intensity to allow for comparison in a sample row for example. The maximum splitting of the $^7F_1$ peak is threefold (2J+1), which can usually still be resolved in high-resolution emission spectra. The transition to the $^7F_2$ reveals a 'hypersensitivity' effect, meaning that its intensity is much more affected by changes in the coordination sphere than predicted for an induced electric dipole transition. Although this 'hypersensitivity' is until yet not fully understood, its effect on the spectrum can be exploited in various ways. For example, the ratio of the $^7F_2/^7F_1$ transition is commonly used as a direct measure for the interaction type of Eu^{3+} with a ligand. A high $^7F_2/^7F_1$-ratio speaks for a stronger binding interaction, like the formation of inner-sphere complexes, where the Eu^{3+} cation is partly stripped from its hydration sphere due to the formation of chemical bonds to a surface for example. In case of the incorporation of Eu^{3+} into a crystal system, the $^7F_2/^7F_1$-ratio can be used as a measure for the degree of disorder of the Eu^{3+} site.

The degeneracy of the individual ^7F_J bands is a result of the crystal-field perturbation effect which, therefore, gives a direct method of probing the local symmetry of the Eu^{3+} site. The according splitting of the J levels can be derived for all space groups from full-rotational group compatibility tables from group theory (s. Table 1 for a cutout).

Table 1: Cutout of splitting of sublevels for Eu^{3+} in dependence on the symmetry class.[120]

Symmetry class	J = 0	J = 1	J = 2	J = 3
Cubic	1	1	2	3
Hexagonal	1	2	3	5
Tetragonal	1	2	4	5
Monoclinic	1	3	5	7
Triclinic	1	3	5	7

When using site-selective excitation, a direct $^7F_0 \rightarrow {}^5D_0$ excitation is performed. An excitation spectrum is then generated by changing the energy wavelength stepwise in a range between around 574 nm to 582 nm and integrating over the measured luminescence intensity. Due to the aforementioned non-degeneracy of both levels, every Eu^{3+} species present in the system will yield one excitation peak from which its resonant energy can be derived and used for selective excitation.

The lifetime (τ) of the Eu^{3+} luminescence can be derived through recording emission spectra with incrementally increasing delay times between laser pulse and signal detection. In aqueous solutions this yields information of the hydration sphere, since H_2O entities can act as luminescence quenchers, due to the coincidence that the energy of the fourth harmonic OH-vibration can bridge the energy gap between emitting and ground state in Eu^{3+}. Therefore, a radiationless transition of the energy can take place, reducing the lifetime of the electronic excitation. *Horrocks and Sudnick*[121] have studied the quantitative relation between the number of H_2O entities in the first coordination sphere of the Eu^{3+} ion and the lifetime of its luminescence and derived a semi-empiric equation, the so called *Horrocks equation* (Eq. (2)).

$$N_{H_2O} = \frac{1.07}{\tau_{Eu^{3+}}[ms]} - 0.62 \quad (2)$$

Accordingly, the lifetimes can be used to distinguish differing sorption species on the surface of zirconia for example, where the hydration sphere differs. Incorporation species can be distinguished readily from sorption species, because of the complete loss of the hydration sphere related to the incorporation. The incorporation of Eu^{3+} ions on differing lattice or non-lattice sites also often yields differing lifetimes of the excited state.

Cm^{3+} Luminescence Spectroscopy

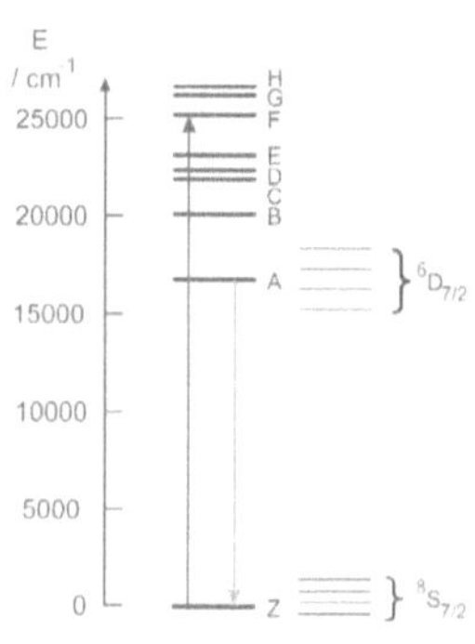

Figure 8: Cm^{3+} term scheme.

The relevant transitions for the absorption and luminescence emission processes of Cm^{3+} are presented in Figure 8.[122] The levels of the Cm^{3+} term scheme are labeled with letters A, B, C… for the excited states with increasing energy and Z for the ground state by convention. The ground state is predominately an $^8S_{7/2}$ state (mixed due to spin-orbit interaction with 17% of $^6P_{7/2}$ and 2% of $^4D_{7/2}$)[123] and the emitting state a $^6D_{7/2}$ state. The strongest absorption can be achieved via the transition from the Z ground state, to the F excited state by UV light (λ_{ex} = 396.6 nm), followed by radiation-less transition to the emitting A state. Both the ground state as well as the emitting state split into four sub-levels, where the ground state splitting is only a few cm^{-1}, but for the emitting A-state it is in the order of 300 – 600 cm^{-1}. Due to the small splitting of the ground state, the Cm^{3+} emission spectra often show only one peak per Cm^{3+} species. Only in very defined systems and when a high resolution can be achieved, the sub-levels can be resolved. However, due to the larger splitting of the excited state, hot-band transitions can often be observed, either as a shoulder or as distinct peaks on the blue-side of the spectrum. With changing complexation of the Cm^{3+} ion, the emission peak shifts to higher or lower wavelengths. This can be explained by a changing crystal field interaction, causing a change of the energy gap between the A and the Z state. Additionally, the nephelauxetic effect, i.e. an expansion of the atomic orbitals in a complex, is observed in dependence of the covalency of the binding interaction. This is a further influence on the energy levels and therefore, the shift of the emission peak.

The shift of the Cm^{3+} emission peak position can be used to obtain a speciation of Cm^{3+} in a multi-species system. In case of sorption processes of Cm^{3+} on a surface for example, an increasing shift is observed with increasing pH due to the transformation from the Cm^{3+} aquo ion to a surface sorbed species. Through spectral deconvolution the single species spectra and their relative fractions can be derived. Examples of Cm^{3+} single species spectra derived through spectral deconvolution are presented in Figure 9. To obtain a correct species distribution, the mismatch of the excitation energy and the energy gap, caused by a change of the ligand field, must be corrected. This is done via the so-called fluorescence intensity factor (FI), derived from the reduction of the luminescence intensity with increasing ligand coordination, the species distribution can be generated.

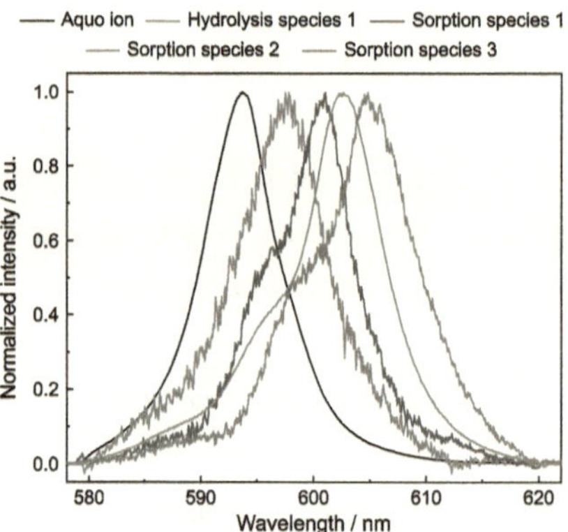

Figure 9: Exemplary Cm^{3+} single species spectra of the sorption on corundum.[86]

The general high sensitivity of the Cm^{3+} emission peak position to the Cm^{3+} environment can be used to distinguish between very similar species, as for example when studying the incorporation into a zirconia host material.

The insight, gained by site-selective excitation of Cm^{3+}, is often inferior to the site-selective Eu^{3+} excitation, since an energy overlap of the individual species is often present, which is amplified by the hot-band transitions.

In analogy to Eu^{3+}, the Cm^{3+} luminescence lifetime is also influenced by the presence of hydration water in the first coordination sphere. The according semi-empirical equation is called *Kimura equation*[124] (Eq. (3)).

$$N_{H_2O} = \frac{0.65}{\tau_{Cm^{3+}}[ms]} - 0.88 \quad (3)$$

This, again, helps distinguishing between species with a difference in the hydration sphere but also between differing incorporation species in solid bulks under the prerequisite that the exchange between the differing species is slower than the luminescence decay. For fast exchange processes, one average lifetime of the participating species is observed.

2.4.2.2. Extended X-ray Absorption Fine-structure Spectroscopy

When X-ray photons hit matter, three different processes can occur, depending on the photon energy as well as the matter the photon is interacting with, i.e. scattering, absorption, or pair production. For absorption of an X-ray photon to take place, the incident energy of the photon must be equivalent to or larger than the binding energy of an electron in the absorbing matter. Due to the high photon energy of X-rays, as compared to laser-light for example, absorption interactions can take place with the core-shell electrons. When the energy of an X-ray photon is below the resonant energy of the according target element, no core shell absorption can take place and mainly scattering processes will reduce the throughput energy. When the energy reaches the specific resonant energy of an absorber, a strong increase of the absorption and therefore decrease of the transition takes place which is the so-called absorption edge. Depending on the shell containing the absorbing electron one can distinguish between the K-, L- or M- absorption edges (Figure 10, top, left). For the analytes studied here, i.e. Zr^{4+} and Y^{3+}, the K-edge energy is the most suitable one with 17.998 keV and 17.038 keV, respectively.

The electron, emitted due to the absorption of the X-ray photon, is back-scattered by the surrounding atoms leading to interference of the several matter waves. Depending on the energy of the incident X-ray photon this leads to constructive or destructive interference at the location of the absorbing atom, causing a change of its absorption coefficient. This results in an oscillating absorption behavior at energies larger than the absorption energy (Figure 10, bottom, left).

Due to this back-scattering effect of the surrounding atoms, information can be gained on the coordination sphere, i.e. the average coordination number of the absorbing element as well as the bond distances to the coordinating atoms around the analyte, typically limited to the first two to three shells. This information is gained by extracting the oscillations from the absorption edge data, usually presented in dependence of the photo-electron wavenumber (Figure 10, top, right) and fitting theoretically calculated individual contributions to the back-scattering of the assumed coordination environment and fitting the parameters mentioned before. Fourier transformation of the oscillations yields spectra in R-space (Figure 10, bottom, right), showing information on the distance of the individual scattering atoms and their abundance, i.e. coordination number. A third fitting parameter is the *Debye-Waller* factor, which is a measure for the divergence within a certain shell. It should be noted that while the distance of the coordinating atoms can be obtained with a typical error of ± 0.01 Å or less, the resulting coordination number is dependent on multiple parameters and, therefore, comes with a rather large uncertainty (± 20%). A more detailed description of the EXAFS method can be found elsewhere.[125]

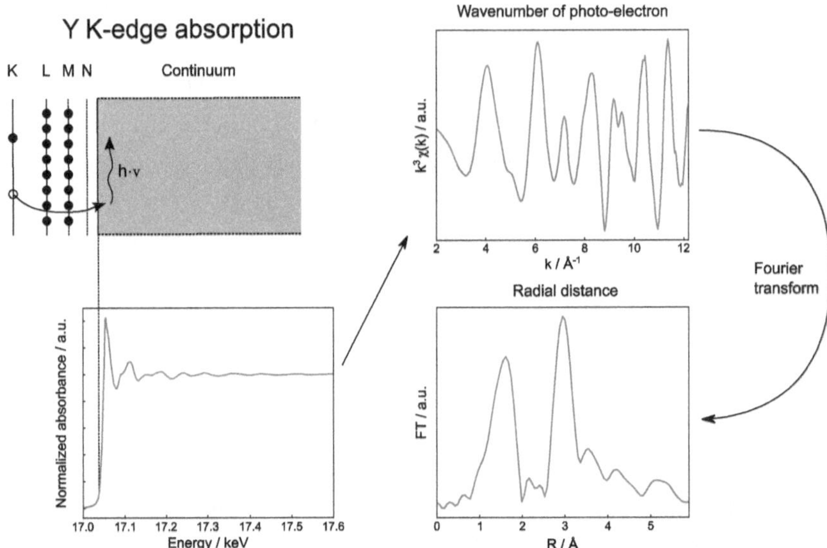

Figure 10: Scheme of electronic transition of Y K-edge absorption (top, left) and relative change of the absorbance with a change of the incident photon energy (bottom, left), transferred to the wavenumber of the photo-electron (top, right) and Fourier-transformed to the radial distance from the absorbing atom (bottom, right).

In the field of HLW repository research, this method can be used to study the nature of a sorption complex, i.e. to which surface groups binding takes place or to get information on the coordination sphere of an incorporated cation.

3. Results and Discussion

In the following chapters the results of this book are presented, including a brief description of the sample preparations and important measurement details. A detailed description of the used materials, sample preparation as well as the measurement setup is given in chapter 5.

3.1. Sorption and Speciation Studies of Eu^{3+} and Cm^{3+} on Zirconia

To enable predictions of possible actinide mobilization in the safety case of water ingress into a HLW repository, the sorption behavior of solubilized and mobilized actinides on the zirconia surface are of high importance. Therefore, the sorption of Cm^{3+} and Eu^{3+} as actinide analogue on pristine monoclinic zirconia was studied in this chapter.

The following results have been published in the Journal of Applied Surface Science.[126]

3.1.1. Mineral Characterization

The commercial ZrO_2 mineral used in this study was pretreated by means of calcination at 1000 °C to remove organic impurities from the surface. These impurities were found to influence the speciation of An(III) at the zirconia interface as described in detail in Eibl et al.[126] The derived pristine material was characterized with various methods.

The PXRD data show the presence of a pure monoclinic phase yielding narrow diffraction peaks with FWHM of approximately 0.19 ° (Figure 11). Using the *Scherrer* equation, a crystallite size of 45 nm could be calculated.

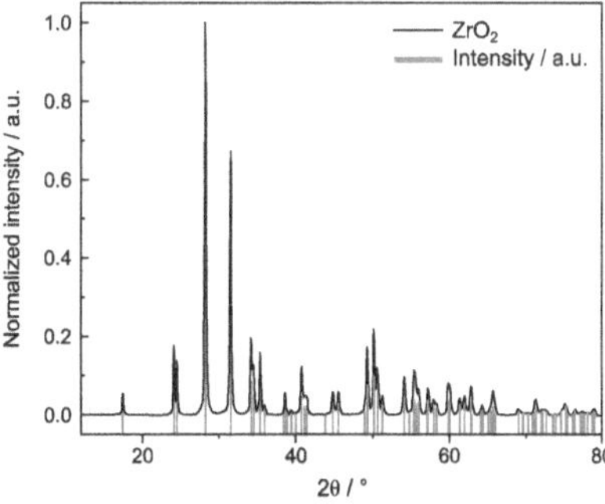

Figure 11: X-ray diffraction pattern of the ZrO_2 used in the sorption and surface studies using Cu-K$_\alpha$ radiation. The reflexes of the monoclinic phase of ZrO_2[127] are indicated with red bars in the figure.

The particle size distribution of the zirconia material, obtained from SEM images, gave a size of 96 ± 32 nm (Figure 12). The particle size differs strongly from the crystallite size, therefore, the observed particles consist of multiple crystallites.

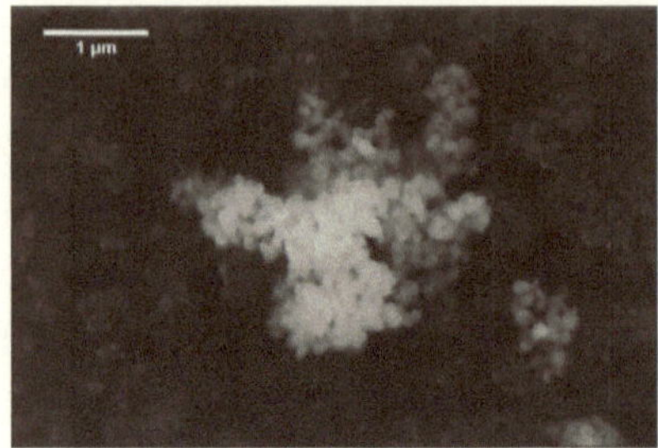
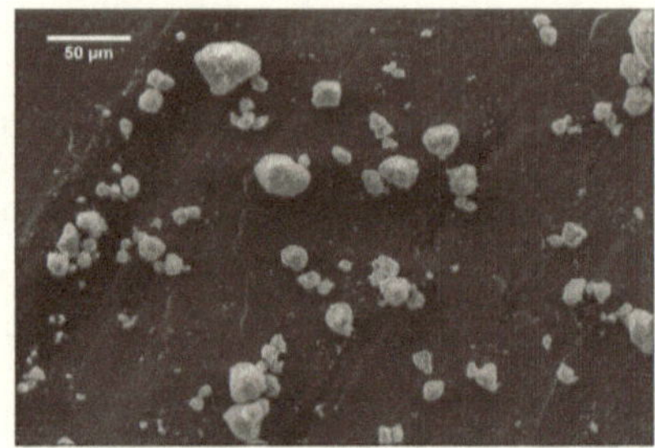

Figure 12: ZrO_2 applied to a sample holder as powder (left) and from suspension (right).

The agglomerate size of the particles was obtained by applying the samples to a sample holder from suspension (Figure 12, right). The resulting agglomerate size is 13 ± 6 µm. This rather high agglomeration tendency of the solid causes low suspension stability.

The BET surface area of this calcined and mortared material is rather small with a value of 5.6 ± 0.1 m^2/g. The micro pore area is 1.26 ± 0.02 m^2/g.

The surface of zirconia was found to have a high affinity towards binding atmospheric CO_2, which can influence the behavior of the materials surface.[128–130] Therefore, total carbon (TC) measurements were performed on the material. The TC present in the material was determined to be 0.13 mg/g. This is assumed to stem from re-adsorbed CO_2 from air, e.g. during cooling after the calcination or during storage at ambient conditions.

The strong influence of atmospheric CO_2 on the surface charge of the ZrO_2 could be observed in zeta-potential measurements (Figure 13). This was studied by two sample rows, one prepared in the absence of atmospheric CO_2 in a glovebox and one at ambient conditions. At inert sample preparation conditions (i.e. N_2 atmosphere), the IEP of ZrO_2 occurs at pH 5.8 which shifts to about 6.9 at ambient conditions. The preparation of all further samples in the study of the surface sorption on zirconia was consequently performed at inert conditions.

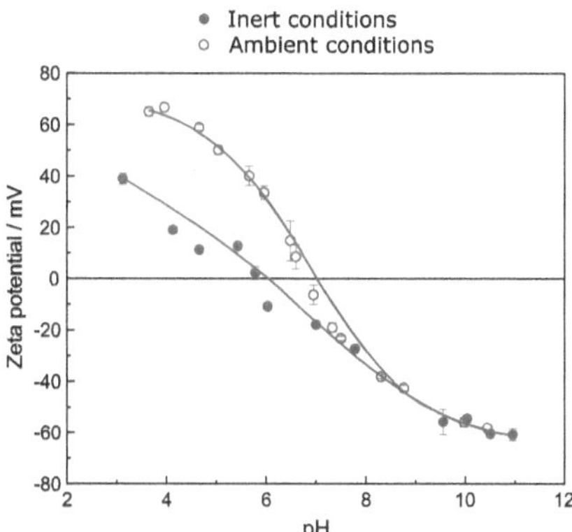

Figure 13: Zeta-potential as a function of pH of 0.5 g/L zirconia with 10 mM NaClO$_4$ as a background electrolyte prepared at inter conditions in a CO$_2$-free glove box (filled symbols) and after exposure to the atmosphere (open symbols).

3.1.2. Eu^{3+} Batch-sorption Experiments on Zirconia

Due to the similar sorption behavior of Eu^{3+} to trivalent actinides, this inactive analogue was used in these studies. The uptake of Eu^{3+} on m-ZrO$_2$ is shown in Figure 14 for varying metal ion and solid concentrations.

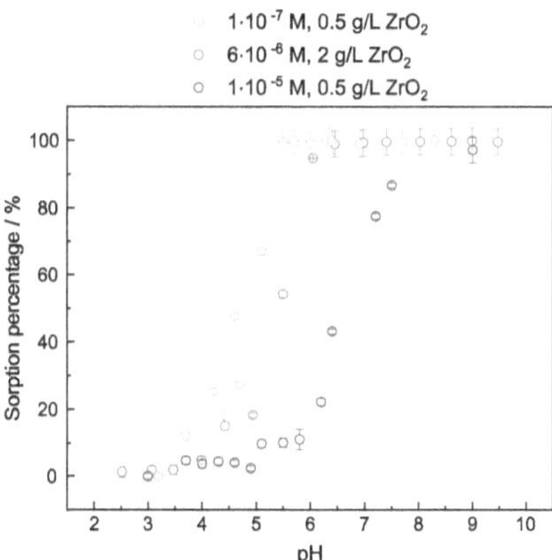

Figure 14: Eu^{3+} batch-sorption curves of ZrO$_2$ in varying metal to mineral ratios with 10 mM NaClO$_4$ background electrolyte.

The sorption curve shows high capability of ZrO$_2$ for the uptake of Eu^{3+}, which is retained on the surface already in the acidic pH range. Sorption can be observed to take place already at a pH of 3.5 while complete sorption is reached at a pH of 5.5 for the lowest metal concentration used. A strong dependence on the metal to solid ratio was, nevertheless, observed. From a low metal to solid ratio of 2·10^{-7} mol/g (1·10^{-7} M on 0.5 g/L) to the intermediate metal to solid ratio of $3 \cdot 10^{-6}$ mol/g ($6 \cdot 10^{-6}$ M on 2 g/L ZrO_2) a slight shift of the sorption curve becomes apparent. A more pronounced effect was observed for the highest metal concentration, i.e. $2 \cdot 10^{-5}$ mol/g ($1 \cdot 10^{-5}$ M on 0.5 g/L), where a strong shift of the adsorption edge is visible. This is influenced by the overall surface coverage. At high metal ion concentrations, various types of surface hydroxyls, i.e. strong and weak sites, can be assumed to participate in the sorption reactions. Precipitation of $Eu(OH)_3$ could be excluded in speciation calculations using the PhreeqC code[131] and the Thermochimie (v. 7) database[132] for the Eu^{3+} hydrolysis constants and solubility product for amorphous $Eu(OH)_3$.

The sorption capacity is dependent on various parameters like the characteristics of the sorption site (i.e. accessibility, reactivity, abundancy), the porosity and pore sizes, the surface roughness and the surface area. The highest influence, however, is the site abundancy, correlating with the surface area, which can be assumed to be the limiting factor here (5.6 ± 0.1 m^2/g). The calcination as pre-treatment for the ZrO_2 could be shown to strongly decrease the surface area of the solid, likely as a result of the increasing crystallite as well as particle size.[126] Nevertheless, the surface sites, calculated based on the surface site density (SSD) of 7.56 sites/nm^2 given in chapter 3.1.4 after calcination is still 3.5 times larger than the number of Eu^{3+} ions present in solution.

When the surface area was sufficient to not decrease the sorption capabilities, complete sorption could be observed to take place at a pH as low as 5.5. This is far below the pH expected for a final repository for HLW, especially when cement is present in the surrounding of the disposed fuel assemblies.

3.1.3. TRLFS Studies of the Cm^{3+} Surface Complexation on Zirconia

Recorded emission spectra for Cm^{3+} sorption on ZrO_2 are presented in Figure 15. The emission spectra show a continuous bathochromic shift as a function of pH, a behavior which is typically noted for the pH-dependent sorption of Cm^{3+} on various mineral surfaces (see e.g. references given in Table 2).

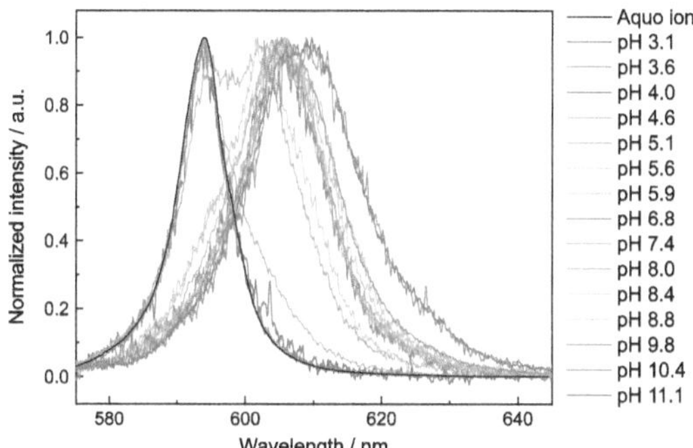

Figure 15: Emission spectra of $5 \cdot 10^{-7}$ M Cm^{3+} on 0.5 g/L ZrO_2 in dependence of the suspension pH (λ_{ex} = 396.6 nm).

Deconvolution of the pH dependent emission spectra was performed to obtain single component spectra of the Cm^{3+} sorption on ZrO_2. Three components were derived with peak maxima of 602.3, 606.2, and 612.5 nm (Figure 16), which are the first reported Cm^{3+} sorption single species spectra on ZrO_2. The peak positions of all three sorption species, especially species 2 and 3, show extremely redshifted emission peaks compared to published Cm^{3+} sorption species on other mineral phases. An overview is given in Table 2.

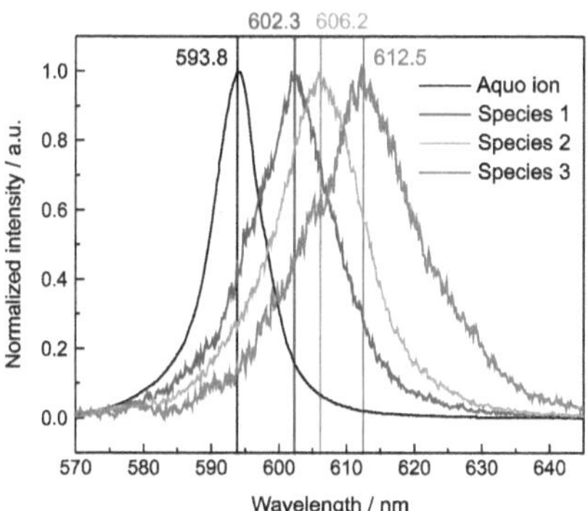

Figure 16: Deconvoluted single species emission spectra for the sorption of Cm^{3+} on ZrO_2.

The strong redshift, i.e. a smaller energy difference between the emitting level and the ground state, of these species means that Cm^{3+} is influenced by a strong ligand field. A strong ligand field may arise from e.g. a high coordination to the surface or an untypically short Cm-O bonding to the surface.

Table 2: Comparison of the emission maxima of Cm^{3+} sorption species on zirconia to published values of various materials.

Sorbent	λ_{max} species 1 in nm / fluorescence lifetime in µs	λ_{max} species 2 in nm / fluorescence lifetime in µs	λ_{max} species 3 in nm / fluorescence lifetime in µs	λ_{max} species 4 in nm / fluorescence lifetime in µs
ZrO_2	602.3 / 90 ± 15	606.2 / 90 ± 15	612.5 / 190 ± 40	- / -
Synthetic kaolinite [84]	598.8 / 109 ± 10	602.6 / 116 ± 8	607.4 / -	610.9 / -
Natural kaolinite [133]	598.8 / 116 ± 13	602.2 / 116 ± 13	606.2 / 132 ± 5	609.4 / 132 ± 5
Corundum (α-Al_2O_3) [134]	600.7 / 110 ± 10	603.0 / 110 ± 10	605.4 / 110 ± 10	- / /
γ-Al_2O_3 [88]	600.6 / 110	602.5 / 110	605.7 / 110	- / -
Bayerite [134]	600.6 / 110 ± 10	603.6 / 110 ± 10	606.7 / 110 ± 10	- / -
Gibbsite [135]	603.5 / 135 ± 7	605.6 / 110	- / -	- / -
Silica colloids [136]	602.3 / 220 ± 14	604.9 / 740 ± 35	- / -	- / -

For a reliable species distribution, it is necessary to account for changes in the absorption maxima of every individual species (i) since the excitation wavelength (λ_{ex}) of 396.6 nm used in these studies correlates to the absorption maximum of the aquo ion only. For that, the relative fluorescence intensity (FI) factors (f_i^{rel}) are used, which can be obtained by the normalization of the single species emission spectra to the overall intensity of the aquo ion. However, this can only be done when the experimental conditions, such as the overall Cm^{3+}-concentration in the sample, remain constant. Due to the very low suspension stability of the ZrO_2 and due to adsorption of zirconia onto cuvette walls and the pH-electrode, such stable conditions could not be ensured in the present work. Therefore, a new approach for determining the FI factors was brought to use.

Known FI factors of published Cm^{3+} species on various minerals under CO_2-free conditions were plotted together and fitted with an exponential curve (Figure 17). The resulting fit follows equation 4 which was used to calculate the FI factors for the zirconia system.

$$f_i^{rel} = 1.2745 \cdot 10^{38} \cdot \exp\left(-\frac{\lambda_{emission}}{6.773}\right) \qquad (4)$$

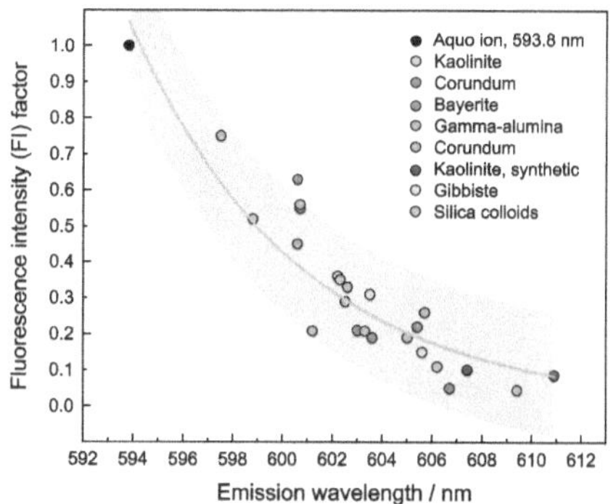

Figure 17: FI factor calibration curve fitted to literature values for kaolinite and synthetic kaolinite[84], corundum [86,134], bayerite[134], γ-alumina[137], gibbsite[135] and silica colloids[136]. An exponential function was used to fit the experimental data. Prediction band (95%) plotted in pink.

With this method, the following FI factors were obtained for all three species (Table 3).

Table 3: Resulting FI factors for the Cm^{3+} species on zirconia, derived using the above formula.

	species 1	species 2	species 3
λ_{max} in nm, FI factor	602.3, 0.32	606.2, 0.16	612.5, 0.01

One should bear in mind that this method is an approximation with rather high uncertainties and needs to be evaluated by experimental results. The error margin can be estimated when calculating the FI factor for the aquo ion using equation 4, which yields a factor of 1.05 instead of 1 by definition.

The FI factor corrected species distribution is plotted in Figure 18, left. The percentage of the aquo ion is at 100% until a pH of about 3.5 where it starts to decrease until a pH of about 5.5 where no aquo ion can be seen in the spectra anymore. A systematic decrease and increase of the different species with increasing pH (Figure 18, right) can be observed over the whole range. The percentage of the aquo ion decreases from pH 3.5 while the first surface sorbed species (species 1) increases, reaching its maximum fraction of 65% at a pH of 5.5. After that, species 2 is dominating over a large pH range from pH 7 to pH 10.5 with a maximum fraction of 100% at pH 8.5. Species 3 starts to appear at a pH of 8.9 and increases above 50% at a pH of 10.5.

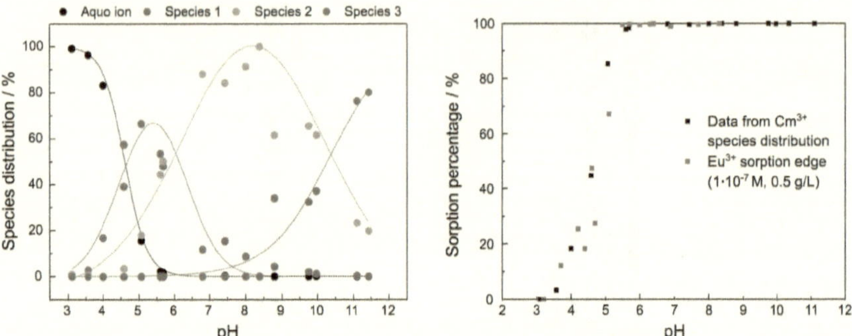

Figure 18: Fluorescence intensity corrected species distributions of Cm^{3+} sorption onto ZrO_2 (left). The dashed lines are a visualization help only. pH dependent sorption behavior of Eu^{3+} on ZrO_2 (red squares) as shown in chapter 3.1.2 in comparison to the Cm^{3+} sorption behavior derived from the species distribution (black squares, right).

When subtracting the sum of all sorption species from the amount of aquo ion left in the system, a pH-edge comprised of only inner-sphere sorption species is obtained as an outer-sphere complex cannot be distinguished in its luminescence behavior from the aquo ion. By comparing this pH-edge with the one derived from Eu^{3+} batch-sorption studies (s. chapter 3.3.1.2), an almost identical shape is obtained (Figure 18, right). Therefore, all sorption species present in the system can be attributed to inner-sphere sorption and no outer-sphere sorption takes place, as a differing edge position would be obtained otherwise.

Due to the high agglomeration tendency of ZrO_2, the lifetimes of the Cm^{3+} sorption on this material yields fairly scattered results (s. Figure 19).

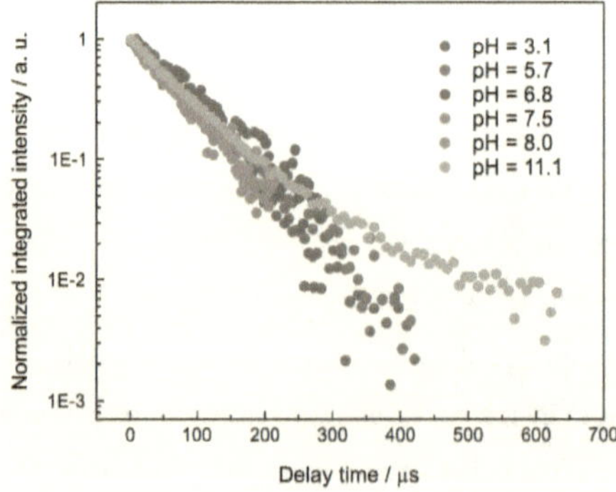

Figure 19: Lifetimes of Cm^{3+} sorption species on ZrO_2 at various selected pH values from about pH 3.1 to 11.1.

Nevertheless, bi-exponential fits of all decay curves yield two lifetimes in addition to the one of the aquo ion, i.e. 90 ± 15 µs which can be assigned to species 1 and 2 due to the species distribution, and 190 ± 40 µs, which correlates with species 3, present at high pH values. The number of hydration water molecules can be estimated to be 6 ± 1 for species 1 and 2 and 3 ± 1 for species 3.

When assuming a preserved overall coordination number of 9, tridentate coordination to the surface must occur for species 1 and species 2. For species 3, a higher surface coordination is to be expected. However, the unusual large redshift of species 3 as well as the low amount of hydration water molecules could be the result of a differing mechanism, such as partial incorporation of Cm^{3+} into the mineral or additional complexation with other ligands, like carbonates. From the total carbon measurements, carbon impurities, most likely, re-adsorbed CO_2 are present on the ZrO_2 surface. Carbonate species are known to form ternary surface complexes with trivalent Eu^{3+} and Cm^{3+} ions resulting in rather long luminescence lifetimes due to the replacement of H_2O/OH^- entities by the carbonate ligand in the metal ion coordination spheres.[129,130] Such an increase of the recorded lifetime due to the presence of carbonates on the mineral surface would lead to an overestimation of the predicted surface coordination based on the number of H_2O/OH^- entities in the Cm^{3+} coordination sphere, and further to an erroneous description of the surface complexation reaction used in this model. However, the lifetimes observed here are shorter than the ones typically reported for Cm^{3+}-carbonate surface species which are generally above 300 μs.[129,130] Further, the samples were prepared in inert atmosphere to avoid carbonate impurities.

The solubility of ZrO_2 is very poor even at pH values between 9 and 12, where species 3 is observed.[36] However, it is known that the surface of zirconia shows a very high reactivity due to a high concentration of unsaturated valences on the surface leading to a strong adsorption of ions and molecules in its environment, such as Cm^{3+} or water.[49] The dissociative adsorption of water leads to an alteration of the surface, resulting in surface layers of very differing properties to the bulk material. This surface alteration effect could be present on the surface yielding a less crystalline product and therefore making it easier to break down deeper layers of the bulk. Consequently, species 3 could be the result of a partial incorporated species into such amorphous surface layers on the zirconia solid.

3.1.4. Surface Complexation Modeling

The Diffuse Double Layer model has been used to describe the sorption process. The SSD was calculated using a crystallographic approach (for details see chapter 5.5.1.5). The obtained set of surface complexation modeling parameters is given in Table 4.

Table 4: Parameter set for surface complexation modeling of Cm^{3+} sorption onto calcined ZrO_2.

Model parameter	Value	Reference
Site density / nm^{-2}	7.56	This work
Protolysis constant 1 ($pK_{a1}°$)	4.0	Blackwell and Carr[138]
Protolysis constant 2 ($pK_{a2}°$)	8.1	Blackwell and Carr[138]
Formation constant for Cm^{3+} species 1 ($\log_{10}K°$)	-3.04	This work
Formation constant for Cm^{3+} species 2 ($\log_{10}K°$)	-9.04	This work
Formation constant for Cm^{3+} species 3 ($\log_{10}K°$)	-22.45	This work

The calculated SSD value is in good agreement with values from Jung and Bell[139], who report a value of 6.2 OH surface groups per nm^2 surface area for a similar, monoclinic zirconia sample.

The FI-corrected species distribution obtained for Cm^{3+} sorption on zirconia was used to obtain the surface complexation constants. For that, the coordination to the surface was assumed to be the difference between the overall coordination number of 9 and the number of coordinating water molecules which were estimated from the obtained luminescence lifetime data. Figure 20, left shows the modeled species distribution (lines) together with the experimental data points (filled symbols). The obtained surface complexation constants were thereafter used to model the resulting Cm^{3+} pH-edge, obtained by subtracting the percentage of adsorbed species from the aquo ion (line in Figure 20, right).

A rather good match between the modeled pH-edge (solid line, Figure 20, right) and the experimental data (filled, grey symbols) was obtained, as well as between the experimental data and the pH-edge derived from the FI-factor corrected Cm^{3+} speciation (filled, black symbols). From the experimentally derived species distribution (Figure 18), it is evident that the edge position is mainly influenced by species 1.

The species distribution clearly shows a discrepancy for species 2 and species 3. For species 2, where a bidentate coordination to the surface was derived from the lifetime data, a clear underestimation of the model can be seen. The opposite is true for species 3, where a pentadentate surface sorption was used in the model, assuming that –OH entities of a hydrolyzed Cm^{3+} cation have a slightly lower luminescence quenching effect than H_2O molecules (non-hydrolyzed Cm^{3+}).[140]

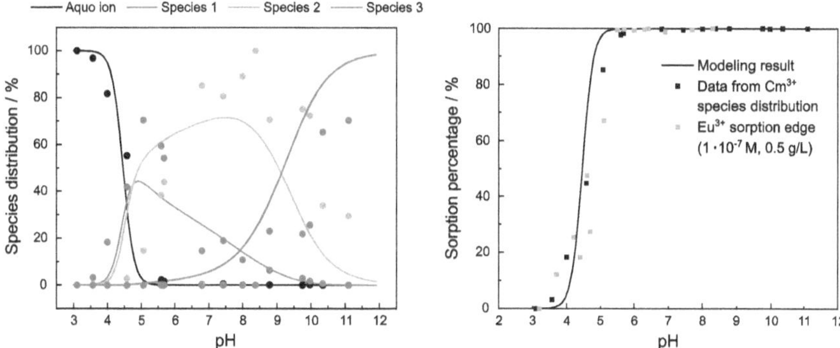

Figure 20: Comparison of the modeled species distribution (lines) to the experimental data (filled symbols) (left) and resulting sorption edge for Cm^{3+} sorption onto ZrO_2 (line) in comparison to the experimental data obtained from the Cm^{3+} species distribution (black squares) and from Eu^{3+} batch-sorption studies (grey squares) (right).

It can be excluded that differences of the surface deprotonation cause the discrepancies between the modeled and the experimental data, as the degree of deprotonation of the participating surface hydroxyl groups has been varied in the model to include every chemically reasonable case. However, since the assumed surface species, for which the formation constants were fitted, are only based on the information on the amount of water molecules in the first hydration sphere and the probable number of surface groups bound in the complex, the speciation is associated with some uncertainties.

Especially species 3 is highly overestimated by the model. As discussed in the previous paragraph, the high denticity of the coordination of Cm^{3+} to the surface might indeed explain the strongly redshifted emission spectra recorded in the laser spectroscopic investigations.

This may be due to an over/underestimation of the surface coordination, however, discrepancies in the actual uptake mechanism and participating ligands may play an important role as well. The modeled Cm^{3+} species is assumed to be an ordinary surface complex coordinating to surface hydroxyls only. Therefore, additional carbonate complexation or partial incorporation of Cm^{3+} into amorphous ZrO_2 bulk may be reasons for the discrepancies observed between the experimental and modeled species distribution. Thus, in order to improve the model, additional information of the Cm^{3+} speciation on ZrO_2 would be required.

For this, future studies providing information of the metal ion coordination sphere, such as X-ray absorption spectroscopy are necessary to access the lacking data required for such a refinement of the surface complexation model. Attempts were made to obtain this information via EXAFS studies of Am^{3+} sorption on ZrO_2. As the concentration of Am^{3+} which can sorb to the ZrO_2 surface is too low for measurements in transmission mode, the EXAFS measurements had to be done in fluorescence mode. Due to a very strong fluorescence of Zr, even at the Am L_I edge which is 5.8 keV higher in energy than the Zr K-edge, no EXAFS signal from the Am could be measured. This problem could be resolved in future studies by the usage of an adsorptive of a lower edge energy than the Zr K-edge, such as a lanthanide (L-edge) or Y (K-edge).

3.1.5. Conclusions of Sorption and Surface Studies of Zirconia

The sorption of the trivalent lanthanide Eu^{3+} was observed to start already in the acidic pH range ~3.5 and reach complete sorption at a pH of 5.5 for low surface coverages. Furthermore, the spectroscopic studies of Cm^{3+} sorption onto zirconia shows that inner-sphere complexation governs this sorption process. This differs from the adsorption reactions on other solid phases of the multi-barrier confinement, such as clay minerals as well as aluminium (hydr)oxides or silicates. In the acidic pH-range, actinide sorption on clay minerals, which are likely to be present in a large abundance in a HLW repository due to the advantageous properties of bentonite as back-fill material as well as possible host clay formations, take place mainly through outer-sphere sorption on the permanent negative surfaces of the clays.[141] Aluminium (hydr)oxides as well as silicates on the other hand show rather low sorption capacities in the acidic pH range and sorption only starts to strongly increase at pH values larger than 5.[84,130,142]

Therefore, zirconia could be of importance for the actinide retardation when a low pH aqueous phase is present. This could be the case due to the formation of iron hydroxides in the proximity of the SNF rods as a result of steel canister corrosion. However, a very similar sorption behavior was observed in previous studies for iron hydroxides, which will be present in larger abundance in a HLW repository.[143]

The species 3 obtained in the speciation studies of Cm^{3+} sorption complexes on zirconia showed a differing behavior from the two other sorption complexes in terms of peak positions as well as lifetimes. As discussed above, this could be the result of a partial incorporation into amorphous surface layers on zirconia. Further research is necessary to fully clarify the nature of this species, however, a partial incorporation species would result in strong retention capabilities of zirconia towards actinides, making it very interesting for future safety considerations.

3.2. Incorporation Studies of Eu^{3+}, Y^{3+}, and Cm^{3+} into Zirconia

To understand the potential retention behavior of the incorporation of actinides into zirconia, a fundamental understanding of doped zirconia phases is crucial. To do so, doped zirconia was prepared by the co-precipitation of hydrous zirconia in the presence of the respective dopant, followed by calcination to obtain a crystalline product (for details see chapter 5.5.2.1). With this method the behavior of various REE dopants, i.e. Y^{3+}, La^{3+}, Ce^{4+}, Pr^{3+}, Nd^{3+}, Sm^{3+}, Eu^{3+} Gd^{3+}, Tb^{3+}, and Lu^{3+}, as well as the actinides Cm^{3+}, Am^{3+}, and Th^{4+} was studied during the making of this book. Detailed studies over a broad range of doping fractions from 500 ppm to 24 mol% were performed using Eu^{3+}. This is due to the advantageous luminescence properties of Eu^{3+}, which gives the opportunity to get a combined bulk structure information via PXRD as well as information on the local structure via TRLFS. Furthermore, EXAFS experiments were used for additional insights into the coordination environment of Zr^{4+} and the dopant. For this, however, Y^{3+} was used as analyte instead of Eu^{3+} due to a more advantageous K-edge energy of 17.0 keV in comparison to the low-energy L_I-edge of Eu (8.1 keV). To also take advantage of the luminescence properties of Eu^{3+} in samples containing other dopants, such as Y^{3+}, co-doping with Eu^{3+} was performed. The Eu^{3+} content in these samples was 0.5 mol%, except for the low Y^{3+} doping row of 0.5 - 2 mol%, where no co-doping took place. For reasons of simplification, all sample compositions are named after the sum of the overall doping concentration and the main crystal phase (m, for monoclinic phase, t, for tetragonal phase or c, for cubic phase) in the sample. The doping fractions of the doped zirconia phases were derived by ICP-MS analysis of the starting solution as well as in the supernatant after the precipitation. This way of determining the Zr^{4+} and Eu^{3+} content in the supernatant is indirect and other methods addressing the dopant concentration in the solid phase will be discussed later in the text.

Some of the following results focusing on the stabilized zirconia phases were published in the Journal of Material Science.[144]

3.2.1. Characterization of Doped Zirconia Phases

PXRD is a key technique to study the phases present in products of the zirconia synthesis. As described above, three zirconia phases were obtained in these studies, i.e. monoclinic ZrO_2 of the space group $P2_1/c$, tetragonal ZrO_2 of the space group $P4_2/nmc$ and cubic ZrO_2, space group $Fm\bar{3}m$. The diffractograms of the three relevant zirconia phases are presented in Figure 21, left. A strong overlap of diffraction peaks of t-ZrO_2 and c-ZrO_2 is present, which can hamper a precise phase analysis. Nevertheless, especially in a row of varying doping fractions, acceptable quantitative data can be obtained, with a relatively large error.

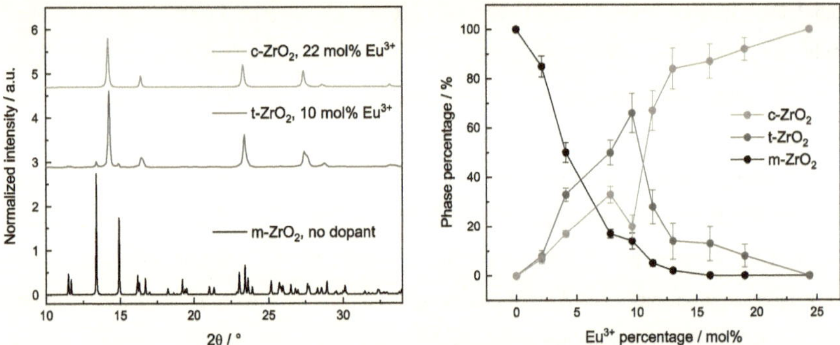

Figure 21: Diffractograms of m-, t- and c-ZrO$_2$ doped with 0, 10, and 22 mol%, respectively, measured at a photon energy of 16.8 keV and plotted with increasing offset (left). Phase fractions of m-, t-, and c-ZrO$_2$ in dependence of the Eu^{3+} doping fraction in a range of 0–24 mol% derived from Rietveld refinement (right).

When performing quantitative *Rietveld* analysis in the Eu^{3+} doping row from 0 – 24 mol%, a systematic phase trend can be observed, as presented in Figure 21, right. A gradual decrease of the initially dominating m-ZrO_2 phase to stabilized zirconia, i.e. tetragonal and cubic zirconia, can be seen with increasing Eu^{3+} doping. The highest fraction of t-ZrO_2 is observed at 10 mol% Eu^{3+} doping, where 66% of the crystal structure is present in the tetragonal polymorph. At higher percentages the cubic phase increases and dominates from about 16 mol% onwards. At 24 mol% Eu^{3+} doping, a pure cubic phase is formed. It should be noted that the doping fractions, necessary to stabilize one or the other polymorph is dependent on the synthesis conditions and cannot be generalized. This is due to the very low cation mobility in zirconia, as will be discussed later.

With increasing doping, the derived lattice parameters of m-ZrO_2 show a slight overall increase, implying that incorporation into the monoclinic phase takes place (Figure 22).

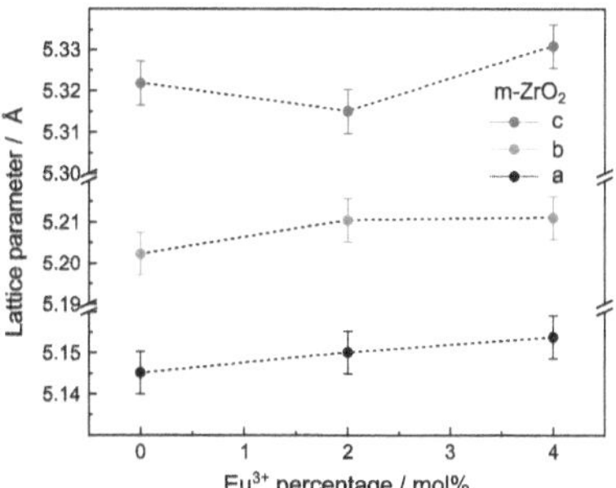

Figure 22: Lattice parameters of monoclinic zirconia in dependence of the europium fraction from 0 to 4 mol%.

The changes of the crystal phases when further increasing the doping fraction are accompanied by a change of the lattice parameters as well, where the a/b parameter of the tetragonal phase increases while the c parameter decreases. (Figure 23, left). In the cubic phase an increase of the lattice parameters a/b/c is observed with increasing doping (Figure 23, right).

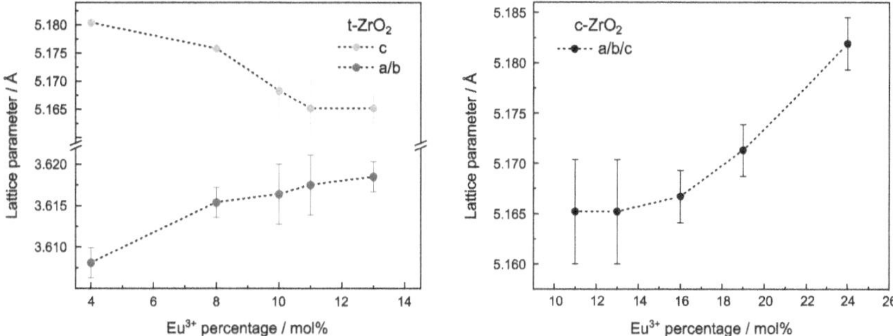

Figure 23: Comparison of the lattice parameters a (= b) and c of the tetragonal zirconia in dependence of the europium fraction from 10 to 13 mol% (left) and lattice parameters of the cubic stabilized zirconia from 11 to 24 mol% Eu^{3+} (right).

The trends of the lattice parameters are not linear here, which is in contradiction to *Vegard's* law, which states that the lattice parameters of a solid solution follows the weighted mean of the two components. Therefore, a linear change of the dopant ratio should result in a linear change of the lattice parameters. This law assumes, however, that the pure phases of the individual components have the same crystal structure and is therefore, only partially applicable here. This deviation from a linear behavior is, however assumed to result from the presence of multiple phases in the system, introducing a significant error margin not only to the phase fractions but also to their lattice parameters.

The lattice parameters can be further used to obtain the doping fractions in the final products in a direct way. As previously stated, the reported final dopant concentrations in the synthetic solid phases are based on ICP-MS analysis of the starting solutions as well as the supernatant after precipitation of hydrous zirconia. However, since it is known that the change of the lattice parameters follows a linear dependence of the doping fraction, the lattice parameters can be used to calculate the doping fraction using equation 5 and 6 in chapter 5.5.2.3.[145,146]

It can be seen that at the low doping range (4 – 8 mol%) and at the high doping range (20 – 24 mol%) a very good match between the added amount of Eu^{3+} and the doping fraction of Eu^{3+} calculated from the lattice parameters is achieved (Figure 24, left). However, at intermediate doping fractions, the doping levels derived from the lattice parameters of the tetragonal phase show a too low doping fraction while the ones derived from the cubic phase lattice parameters are too high. When comparing the trend of the lattice parameters from Figure 23 with these results, the discrepancy in the dopant fractions are most likely due to the large error in the derived lattice parameters rather than a deviation of the actual composition of the solid phases. The inaccuracy of the lattice parameters results from the strong overlap of the diffraction peaks of the t- and c-ZrO_2 phases.

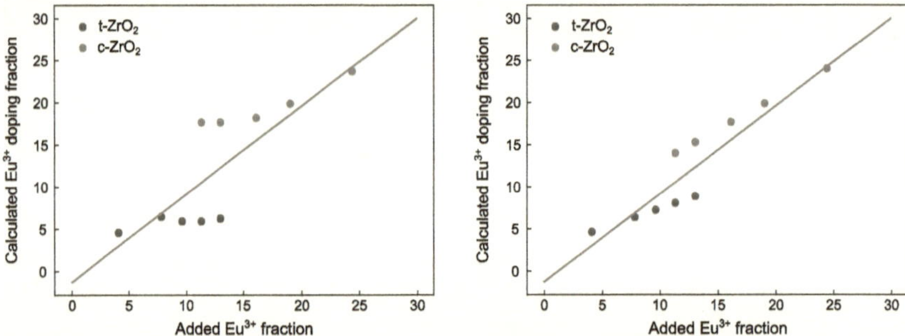

Figure 24: Eu^{3+} doping fraction calculated from the lattice parameters of t- (blue dots) and c-ZrO_2 (red dots) compared to the line of identity (blue line) (left) and values calculated when deriving lattice parameters from an extrapolation from the doping fractions 8 and 10 mol% for t-ZrO_2 (blue dots) and extrapolating from fractions 16, 19 and 24 mol% for c-ZrO_2 (red dots, right).

When extrapolating a linear increase of the lattice parameters a/b of the tetragonal phase and a linear decrease for c, as postulated by *Vegard's* law, for the doping fractions of 4 and 8 mol%, where the lattice parameters have the smallest margin of error, a tentative correction for the inaccurate values can be made. The same can be done for the cubic phase, where a linear extrapolation from the fractions 16, 19, and 24 mol% was made. When calculating the doping fractions based on the hereby derived lattice parameters a much better agreement of the expected doping fractions can be achieved (Figure 24, right). As there are no indications for the opposite, the doping fractions in the final product can be assumed to correlate with the removal of Zr^{4+} and Eu^{3+} from solution, as determined with ICP-MS, and, therefore, with the amount of Eu^{3+} added in the co-precipitation synthesis.

The synthesis conditions used for the preparation of doped zirconia phases are critical, when trying to obtain reproducible phase compositions. Extreme effects on the resulting product were described when using differing precipitation pH-values,[147,148] differing synthesis temperatures[149] or even differing zirconium and dopant-ion sources.[150] The influence of differing dopants was tested in the current study to assess the differences in stabilization as a function of the dopant cation radius.

When varying the dopant cation under constant conditions, surprisingly little changes in the diffractograms (Figure 25, left) and the phase fractions are observed (Figure 25, right), even though the cation radius differs by 9% from Lu^{3+} to Eu^{3+}. Especially in the row from Y^{3+} via Gd^{3+} to Eu^{3+} a very constant ratio of m-ZrO_2 to stabilized ZrO_2 can be seen.

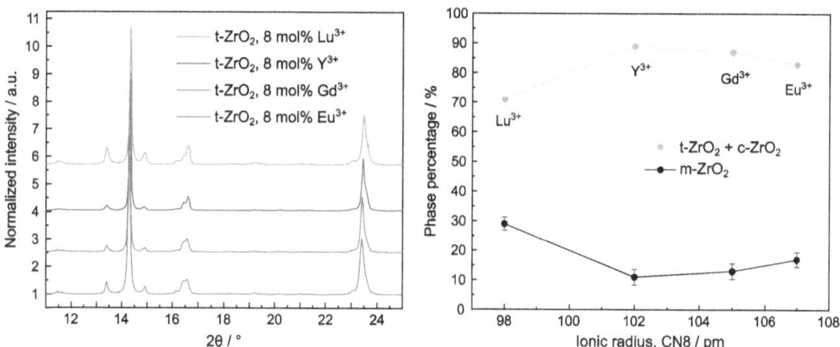

Figure 25: Diffractogram of pristine m-ZrO_2 and Eu^{3+} stabilized t- and c-ZrO_2, measured at a photon energy of 16.8 keV and plotted with increasing offset (left). Effect of differing stabilizing dopants on the phase percentage of stabilized phases (t- and c-ZrO_2) vs. m-ZrO_2, plotted over the ionic radii of the cations in eight-fold coordination (right).[31]

When comparing a row of Y^{3+} doped ZrO_2 of doping fractions from 18 to 26 mol% with the according Eu^{3+} row seen above (s. Figure 21), a very good match between the phase compositions is obtained (Figure 26). The *Rietveld* analysis of Y^{3+}/Eu^{3+} co-doped samples, however, show that in contrast to the purely Eu^{3+}-doped samples, the monoclinic phase persists as a very minor amount up to concentrations of 22 mol%. From thereon, an almost pure cubic phase can be observed. Its percentage increases to 100% for the highest doping percentage, i.e. 26 mol%. Between 18 and 24 mol% a small amount of the tetragonal phase persists. Due to the similarities of the stabilization of the zirconia structure for the set of Eu^{3+} and Y^{3+}/Eu^{3+} co-doped samples, it is assumed that Y^{3+} and Eu^{3+} can be used as analogues for one another which is exploited in the following chapters, where in some cases parallel rows of Eu^{3+} and Y^{3+}/Eu^{3+} co-doped zirconia were studied with different methods.

In all samples, increasing doping fractions lead to a broadening of the diffraction peaks (not shown here) which can have different origins. The increasing amount of the highly oversized dopant cation may introduce an increasing disorder of the bulk structure, resulting in the observed broadening of the diffraction peaks.[151] Generally, however, peak broadening in powder diffraction is mainly the result of small crystallite sizes below 100 nm, reducing the degree of long-range order.

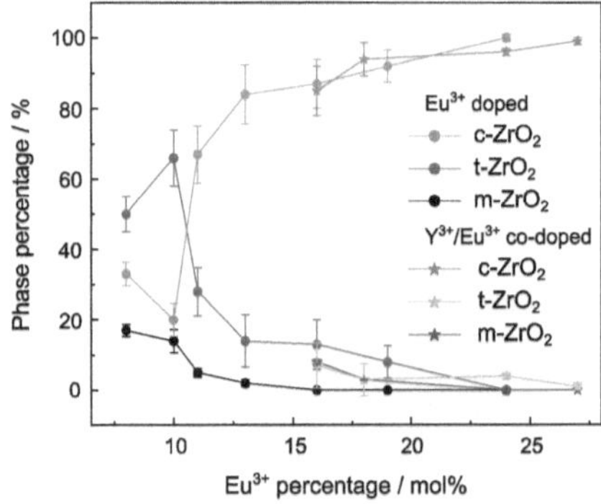

Figure 26: Phase fractions of m-, t-, and c-ZrO_2 in dependence of the Eu^{3+} percentage (8 – 24 mol%) or Y^{3+} percentage (16 – 26 mol%).

When studying the crystallite sizes derived from the PXRD pattern via the *Scherrer*[119] equation (Eq. 1), a decrease with increasing doping fractions can be seen (Table 5). Furthermore, larger crystallite sizes are obtained consistently for Y^{3+}/Eu^{3+} co-doping over the whole compositional range. This is likely an effect of the cation radius, which is larger for Eu^{3+} in comparison to Y^{3+}, resulting in a larger strain and stronger lattice distortion in the samples with a direct influence on the maximum crystallite size.

Table 5: Summary of the main diffraction peak position and the resulting crystallite size derived from Scherer's equation (K = 1) for samples of Eu^{3+} doping of 8 – 24 mol% and for Y^{3+}, Eu^{3+} co-doped zirconia with doping fractions of 16 – 26 mol%. *FWHM is corrected to subtract peak broadening from the setup via a Si reference sample.

Sample	Diffraction peak position / ° 2Θ	FWHM* / 2Θ	Crystallite size / nm
Eu^{3+}			
t-ZrO$_2$, 8 mol%	30.091 ± 0.005	0.269 ± 0.01	34 ± 1
t-ZrO$_2$, 10 mol%	30.091 ± 0.005	0.269 ± 0.01	34 ± 1
t/c-ZrO$_2$, 11 mol%	30.094 ± 0.005	0.290 ± 0.01	31 ± 1
t/c-ZrO$_2$, 13 mol%	30.071 ± 0.005	0.284 ± 0.01	32 ± 1
c-ZrO$_2$, 16 mol%	30.065 ± 0.005	0.300 ± 0.01	31 ± 1
c-ZrO$_2$, 19 mol%	30.025 ± 0.005	0.340 ± 0.01	27 ± 1
c-ZrO$_2$, 24 mol%	29.984 ± 0.005	0.433 ± 0.01	21 ± 1
Y^{3+}/Eu^{3+}			
c-ZrO$_2$, 16 mol%	30.083 ± 0.005	0.103 ± 0.01	88 ± 1
c-ZrO$_2$, 18 mol%	30.097 ± 0.005	0.106 ± 0.01	86 ± 1
c-ZrO$_2$, 24 mol%	29.992 ± 0.005	0.136 ± 0.01	67 ± 1
c-ZrO$_2$, 26 mol%	29.984 ± 0.005	0.144 ± 0.01	63 ± 1

The TEM images of non-stabilized (Figure 27, left) and stabilized (Figure 27, middle and right) zirconia show agglomerates of small particles with a heterogeneous morphology.

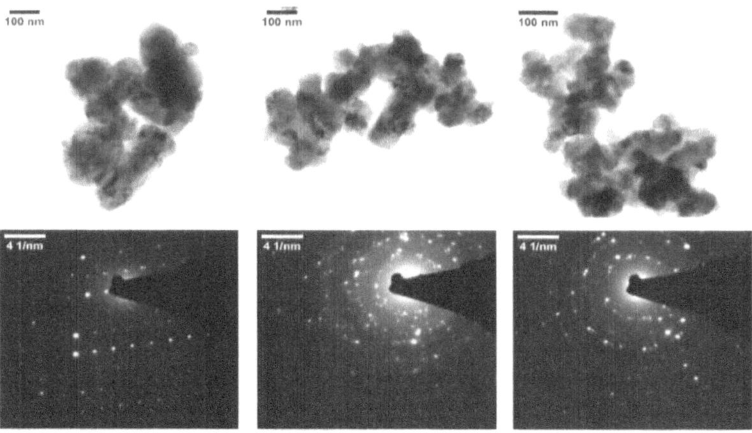

Figure 27: TEM images (top) and electron diffraction images (bottom) of non-stabilized m-ZrO$_2$ (left), t-ZrO$_2$ doped with 8 mol% Eu^{3+} (middle) and c-ZrO$_2$ doped with 22 mol% Eu^{3+} (right).

The images and the according electron diffraction (ED) images support the trend of smaller crystallite sizes at higher doping fractions, since non-doped zirconia shows larger crystallites and the reciprocal lattice of the monoclinic phase in ED (Figure 27). When increasing the doping, the ED images show powder like rings implying multiple crystallites within the beam spot.

SEM images of the polished samples were taken and combined with energy dispersive X-ray spectroscopy (EDX) to investigate the distribution of the dopant ion within the zirconia matrix. Based on the results of the elemental mapping, a homogeneous distribution of the dopant within the zirconia matrix could be confirmed (see Figure 28).

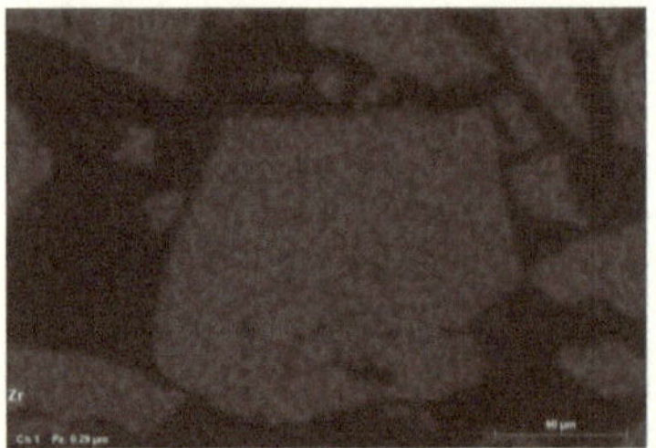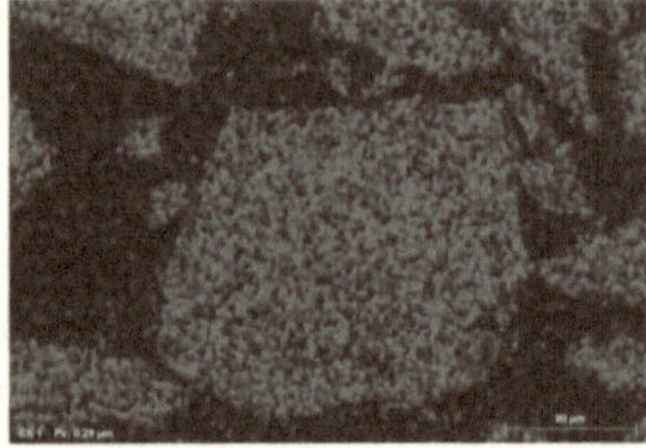

Figure 28: SEM-EDX image of embedded and polished zirconia sample with a Eu^{3+} doping percentage of 30 mol%, showing the distribution of Zr (left, blue) and of Eu (right, red).

3.2.2. EXAFS Study of Trivalent Cation Doping of Zirconia

For a detailed understanding on what structural changes take place on the molecular level, in relation to the changes of the bulk structure discussed above, EXAFS investigations were performed.

The sample compositions for the EXAFS studies were chosen to yield predominantly pure phase compositions, i.e. in the low dopant range from 0.5 – 2 mol% where the monoclinic zirconia phase prevails, and the 16 – 26 mol% region where the cubic structure dominates. In Figure 29 a comparison of the EXAFS (left) and their Fourier transform (right) for the Zr K-edge (top) and the Y K-edge (bottom) is presented for the monoclinic crystal structure. All samples in the range from 0.5 – 2 mol% doping yield a very similar EXAFS with only subtle differences.

The fitting of the m-ZrO_2 structure holds the problem that all seven oxygens, coordinating every Zr^{4+} cation have differing distances from it. Further, the twelve Zr^{4+} cations coordinating a center Zr^{4+} cation are located at eight differing crystallographic distances.

Since fitting every single of these distances would not yield a reasonable fit, one average Zr-O shell and three Zr-Zr shells were used, which yields the best fitting results (s. Table 6). A comparison of the fitting results and the crystallographic data can be found in Table A4.

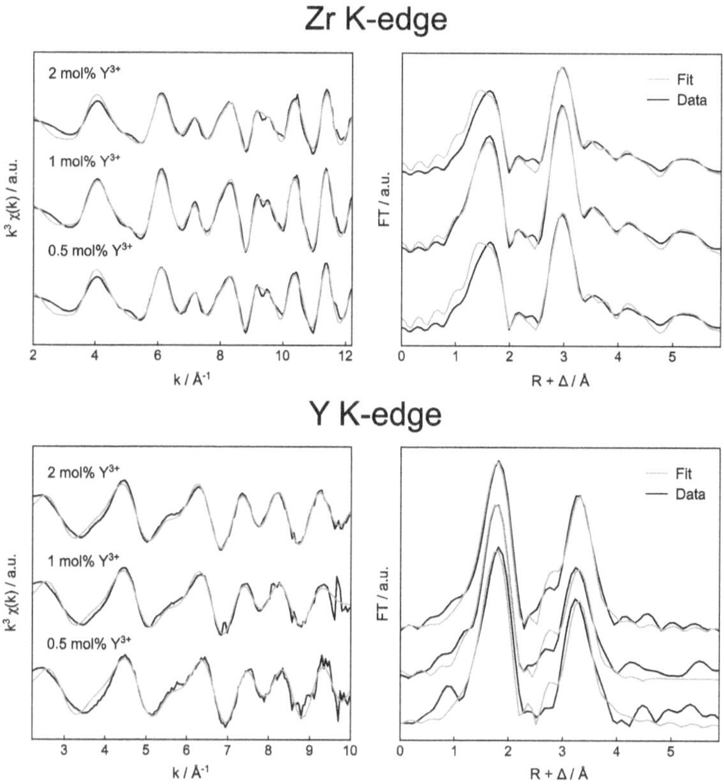

Figure 29: Experimental (black) and fitted (red) k^3-weighted EXAFS data (left) and FT (right) of Y^{3+} doped zirconia with doping levels of 0.5 - 2 mol% at the Zr K-edge (top) and Y K-edge (bottom).

The overall data quality of the Zr-edge data is considerably better, leading to a larger evaluable k-range (2 - 12.2 Å⁻¹) as compared to the Y-edge (2.2 - 10 Å⁻¹). It is evident from the k-space data, that the Y-data, especially for the lower doping samples of 0.5 and 1 mol%, shows already strong artefacts at ~ 9.8 Å⁻¹. The differences in the data quality are caused by the rather obvious differences in the analyte concentrations present in the samples.

The resulting fitting values (Table 6) confirm the very consistent Zr as well as the Y coordination shells over this doping range from 0.5 - 2 mol%.

Table 6: Structural parameters derived from the k^3-weighted EXAFS spectra for the Zr and the Y K-edge. Fixed CN are marked with an asterisk (*). Fitting range is 2 – 10.5 Å^{-1}, R: radial distance, error ± 0.01 Å, CN: coordination number, error ± 20%, σ^2: Debye-Waller factor, error ± 0.0005 Å, amplitude reduction factor (S_0^2) = 1.0.

Y^{3+} content / mol%	Path	Zr K-edge			R-value / %	Y K-edge			R-value / %
		R / Å	CN	σ^2 / Å^2		R / Å	CN	σ^2 / Å^2	
0.5	abs-O	2.14	7	0.015	5.7	2.26	7	0.0095	10.2
	abs-Zr	3.47	7	0.008		3.45	7	0.0116	
	abs-Zr	3.70	4	0.009		3.58	4	0.0050	
	abs-Zr	3.95	1	0.003		3.90	1	0.0050	
	abs-O	4.68	8	0.016		-	-	-	
	abs-O	4.98	6	0.003		-	-	-	
	abs-O	5.70	9	0.005		-	-	-	
	abs-O	6.00	6	0.002		-	-	-	
1	abs-O	2.14	7	0.009	2.6	2.27	7	0.0101	9.2
	abs-Zr	3.45	7	0.006		3.49	7	0.0073	
	abs-Zr	3.62	4	0.012		3.65	4	0.0052	
	abs-Zr	3.97	1	0.003		3.97	1	0.0052	
	abs-O	4.69	8	0.010		-	-	-	
	abs-O	4.96	7	0.001		-	-	-	
	abs-O	5.68	10	0.001		-	-	-	
	abs-O	6.02	7	0.001		-	-	-	
2	abs-O	2.14	7	0.016	6.5	2.28	7	0.0104	3.5
	abs-Zr	3.47	7	0.009		3.50	7	0.0060	
	abs-Zr	3.70	4	0.009		3.64	4	0.0022	
	abs-Zr	3.94	1	0.004		3.96	1	0.0126	
	abs-O	4.68	6	0.010		-	-	-	
	abs-O	4.96	8	0.006		-	-	-	
	abs-O	5.70	10	0.006		-	-	-	
	abs-O	6.00	6	0.002		-	-	-	

The Y-O distance is larger than the Zr-O one, as expected due to the larger ionic radius of the former cation. However, in the Y-Zr shell, only minor differences to the Zr-Zr shell are discernable. The Y-edge data of the 0.5 mol% sample shows the largest deviation to all other data sets (i.e. for 0.5 - 2 mol% for the Zr-edge data and 1 and 2 mol% for Y-edge). This is assumed to result from the rather scattered data resulting from the low amount of analyte, rather than from actual deviations in the structure. Differences at the Zr-edge from the fit and data can be observed for the first coordination shell, especially in 0.5 and 2 mol% Y^{3+} doped ZrO_2. This is assumed to be a result of the background subtraction.

In Figure 30 a comparison of the EXAFS (left) and their Fourier transform (right) for the Zr K-edge (top) and the Y K-edge (bottom) is presented for the high doping range from 16 - 26 mol%, where a mostly cubic phase is present. As seen before in the low doping range, also here, despite of the much larger amount of the oversized dopant, minor differences can be seen in the Zr^{4+} and Y^{3+} coordination spheres over the doping range.

Zr K-edge

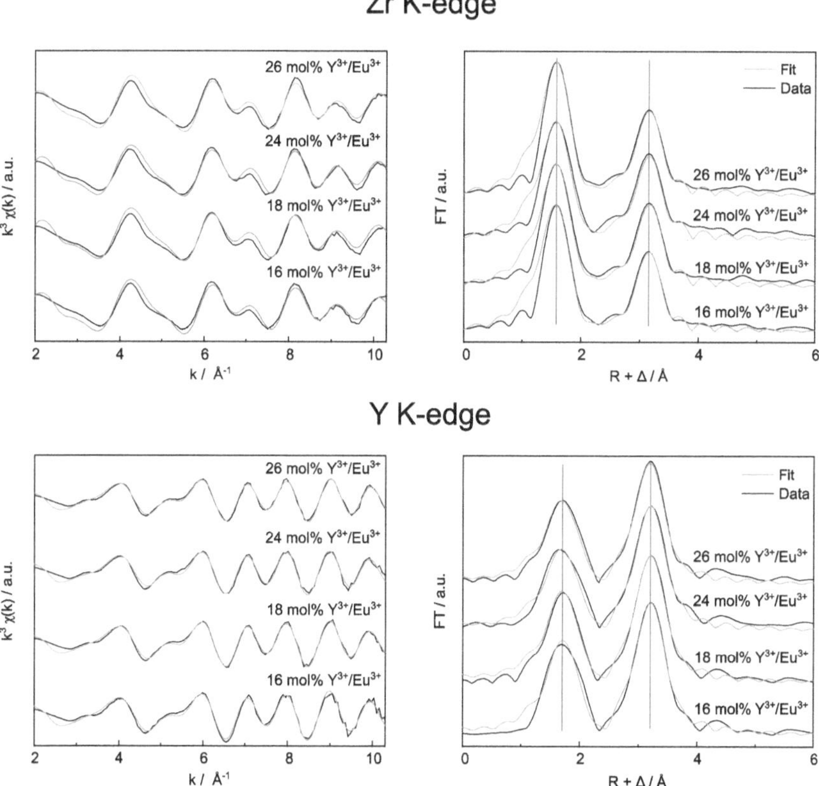

Figure 30: Experimental (black) and fitted (red) k^3-weighted EXAFS data (left) and FT (right) of Y^{3+} doped zirconia with doping levels of 16 - 26 mol% at the Zr K-edge (top) and Y K-edge (bottom). Red lines are a visualization help only.

As the introduction of the trivalent dopant into the ZrO_2 crystal structure is known to result in the formation of oxygen vacancies, a decrease of the M-O (M = Zr, Y) coordination number would be expected with increasing dopant concentration. Therefore, an attempt was made to obtain a fit for the coordination numbers (CN). The CN fitting of the absorber-O path (abs-O) did not yield a reasonable trend (s. Table A1) which is why the CN was fixed to 8 in the final fit. The same is true for the abs-cation (abs-M) CN, which was fixed to 12 (s. Table A2 and Table A3). Indeed, no clear trend can be derived from the obtained values. This is likely due to the larger error associated with the CN, making small variation difficult to quantify. In addition, no splitting was observed for the Zr-O shell which is often described for the tetragonal zirconia phases.[152,153]

Due to the larger bulk symmetry, multiple-scattering (MS) paths were included in the fits here. The obtained average Zr-O distance is very constant over the whole high doping range where a slight increase from 2.14 to 2.15 Å can be observed from 16 to 26 mol% (Table 7).

Table 7: Structural parameters derived from the k^3-weighted EXAFS spectra for the Zr and the Y K-edge. Fixed CN are marked with an asterisk (*). Fitting range is 2 – 10.5 Å^{-1}, R: radial distance, error ± 0.01 Å, CN: coordination number, error ± 20%, σ^2: Debye-Waller factor, error ± 0.0005 Å, amplitude reduction factor (S_0^2) = 1.0.

Y^{3+} content / mol%	Path	Zr K–edge			R–value / %	Y K–edge			R–value / %
		R / Å	CN	σ^2 / Å^2		R / Å	CN	σ^2 / Å^2	
16	abs–O	2.141	8*	0.012	8.18	2.321	8*	0.010	6.86
	abs–M	3.550	12*	0.014		3.612	12*	0.009	
	abs–O–M	3.957	48*	0.019		4.171	48*	0.008	
	abs–O	4.331	24*	0.041		4.503	24*	0.033	
18	abs–O	2.146	8*	0.013	7.37	2.316	8*	0.011	6.13
	abs–M	3.557	12*	0.014		3.607	12*	0.010	
	abs–O–M	3.935	48*	0.024		4.157	48*	0.005	
	abs–O	4.352	24*	0.044		4.488	24*	0.036	
24	abs–O	2.153	8*	0.013	8.86	2.317	8*	0.012	6.20
	abs–M	3.556	12*	0.013		3.605	12*	0.010	
	abs–O–M	3.873	48*	0.027		4.144	48*	0.012	
	abs–O	4.382	24*	0.048		4.476	24*	0.044	
26	abs–O	2.147	8*	0.011	6.74	2.313	8*	0.012	6.02
	abs–M	3.550	12*	0.013		3.605	12*	0.010	
	abs–O–M	3.935	48*	0.026		4.144	48*	0.012	
	abs–O	4.338	24*	0.049		4.476	24*	0.044	

The Zr-cation distance gives a constant value of 3.55 Å varying only within margin of error. A comparison of the derived atomic distances with crystallographic data is presented in Table A4.

The Y-O distance as well as the Y-M distance are constant within margin of error over the whole doping range with values of 2.32 Å and 3.61 Å, respectively, and therefore, larger than for the corresponding Zr shells. This translates to the shells at higher radial distance, i.e. the MS path of M-O-M as well as the second Y-O shell where also larger values are obtained in comparison to the corresponding Zr paths.

It is striking that despite the rather large difference in the Zr–O and Y–O distance (~8%) the Zr–cation and Y–cation distances are very similar (difference of ~1.5%). The changes introduced by the oversized dopant are therefore, mostly limited to the first coordination sphere. This observation was explained by Li et al.[101] with the deformability of oxygen ligands as well as with the continuum elasticity theory, where a quadratic decay of the radial displacement is predicted.

The derived abs-O and abs-M distances agree very well with existing data for Y-doped ZrO_2 samples of various compositions.[102,154] They also confirm the trend reported in multiple studies, where hardly any change in the host or dopant distances to the coordinating oxygen atoms can be deduced despite the linear increase of the unit cell parameters derived from XRD data.[102,155]

Comparing these results to the ones from the monoclinic sample row (0.5 – 2 mol%) the Zr^{4+} environment revealed to be very similar in both cases. The average Zr-O distance is 2.14 Å in m-ZrO_2 and 2.14 ± 0.01 Å in c-ZrO_2. The Zr-M shell results in 3.58 ± 0.02 Å and 3.55 ± 0.01 Å for m-ZrO_2 and c-ZrO_2, respectively. Differences are, however, present in the Y^{3+} environment where the average Y-O distance is 2.27 ± 0.01 Å in m-ZrO_2 and 2.32 ± 0.01 Å in c-ZrO_2 and the average Y-M distance equals to 3.57 ± 0.04 Å in m-ZrO_2 and 3.61 ± 0.01 Å in c-ZrO_2. Clearly, the Y environment is more compressed when incorporated into m-ZrO_2 which has a larger site volume as compared to stabilized zirconia. However, in both cases the effect is dampened in the second coordination sphere.

Overall, no changes can be observed within the m-ZrO_2 and the stabilized ZrO_2 phases over the doping range studied here, indicating that no additional phases or species are present in the system.

While EXAFS can yield valuable information on the average coordination sphere of an analyte in a structure, the presence of multiple species is not always detectable. Reasons for that can be that these species have a very similar environment, and, therefore an average between two similar species is obtained. Another reason can be that only a small fraction of a certain species is present, causing its signal to be covered up by the dominant species. Lastly, destructive interference of the EXAFS signal caused by similar analyte environments can obliterate the signal of both species.

To study the potential presence of multiple species in more detail, TRLFS studies were performed.

3.2.3. TRLFS Studies of Doped Zirconia Phases

Depending on the object of interest, TRLFS can yield additional insights into the system under investigation, especially when multiple species are present in the same sample. To gain further insight into the incorporation process of trivalent metal cations in zirconia, TRLFS investigations of Eu^{3+}-doped, Y^{3+}/Eu^{3+} co-doped, and Cm^{3+} doped ZrO_2 samples were conducted.

3.2.3.1. Eu^{3+} UV-TRLFS Studies of the Local Dopant Environment

UV-excitation ($^5L_6 \leftarrow {}^7F_0$)

Some selected emission spectra of Eu^{3+} doped ZrO_2 in the lower doping range from 500 ppm to 4 mol%, obtained by UV-excitation, are presented in Figure 31.

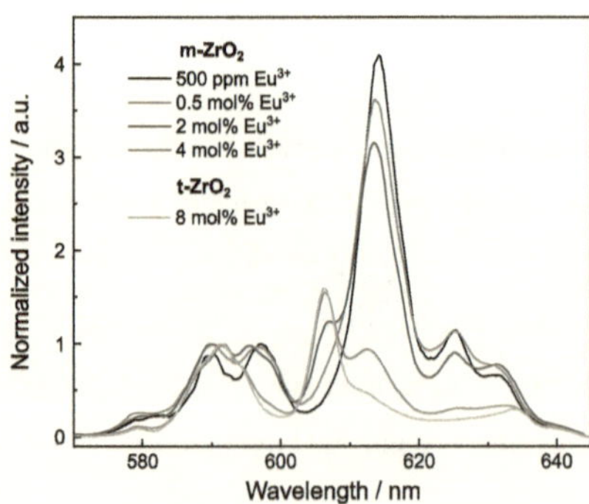

Figure 31: Emission spectra of m-ZrO_2 doped with Eu^{3+} in fractions from 500 ppm – 4 mol%.

It can be seen how the gradual increase of stabilized zirconia phases (t- and c-ZrO_2) translates to the emission spectra by means of a decrease of the main peak of the monoclinic phase at

614 nm accompanied by a shift to lower wavelengths. At 4 mol% doping, where the monoclinic phase makes up only 50% of the phase distribution, the peak intensity of the main peak of m-ZrO_2 is already lower than the one of t- and c-ZrO_2. At 8 mol% this peak is almost gone. A close look at the $^7F_1$ band, however, also shows that more than one species must be present even at 500 ppm doping, as at least four shoulders can be observed here.

The emission spectra of the sample row with 8 mol% to 24 mol% Eu^{3+} doping are presented in Figure 32, left. The peaks deriving from m-ZrO_2 are not present anymore at this doping level. Further, the emission spectra show rather broad emission peaks, especially in the $^5D_0$ → $^7F_1$ transition, hindering a clear determination of the Stark level splitting and, therefore, clear information on the site-symmetry of the Eu^{3+} cation. The $^5D_0$ → $^7F_2$ transition exhibits only two distinct peaks at 606.4 nm and 634.0 nm with additional unresolved broad intermediate shoulders. It becomes apparent, that the emission of both stabilized phases, the tetragonal and the cubic zirconia, yield indistinguishable spectra.

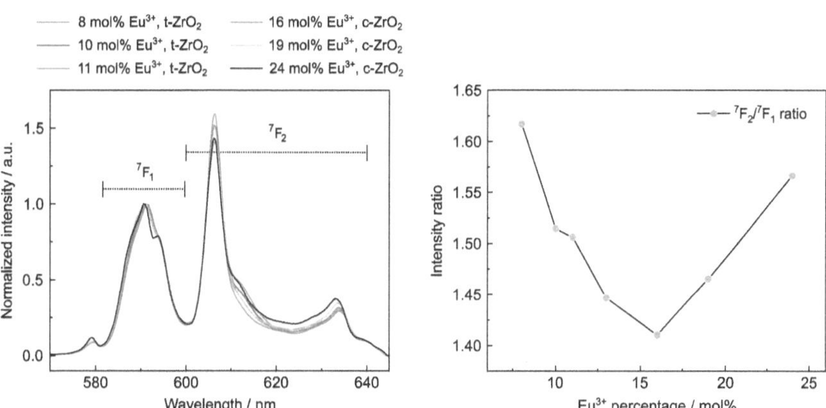

Figure 32: Selected luminescence emission spectra of Eu^{3+} doped zirconia (8 - 24 mol%) at UV-excitation (λ_{ex} = 394 nm) (left), intensity ratio of the $^5D_0$ → $^7F_2/^7F_1$ transitions (right).

When studying the changes of the $^5D_0$ → $^7F_2/^7F_1$ intensity ratio, which gives information on changes in the local symmetry, with respect to the dopant percentage, a certain trend can be derived (Figure 32, right). With increasing dopant concentrations from 8 mol% up to 11 mol% a decreasing $^7F_2/^7F_1$ ratio can be seen, which can be interpreted as relative increase of the Eu^{3+} site-symmetry. This can be partly explained by the decrease of the remaining monoclinic phase until 11 mol%. At higher percentages, the $^7F_2/^7F_1$-ratio decreases further until the minimum at 16 mol%, which is where about 90% of the zirconia is cubic. From thereon, the intensity ratio shows increasing behavior with increasing dopant fractions and almost reaches the same level of asymmetry at 24 mol% as observed at 8 mol%.

From a mere symmetry point of view, this behavior is not foreseen, as the transition from a tetragonal to cubic environment should result in an increase of the relative symmetry and subsequent decrease of the $^7F_2/^7F_1$ ratio. However, the increasing concentration of dopants in the lattice has two effects on the Eu^{3+} sites. Firstly, the high concentration of Eu^{3+} in the undersized host lattice is likely to cause large distortions of the site symmetry and, therefore, also affect Eu^{3+} cations in proximity. Secondly, the high concentration of Eu^{3+} creates a high concentration of oxygen vacancies, changing the coordination environment of all components in the system.

To obtain more insights into M^{3+} doped ZrO_2, especially where multiple species are present at once or where the resolution of room temperature UV-excitation measurements is not good enough, site-selective TRLFS studies (at T < 10 K) are presented in the following chapter. From the UV-TRLFS studies of Eu^{3+} doped ZrO_2 it becomes apparent that the processes occurring when using low amounts of dopant differ strongly from highly doped ZrO_2. Therefore, this chapter will be divided into studies of low doping phases (3.2.3.2), targeting dopant fractions from 0 – 7 mol% and studies of highly doped zirconia phases (3.2.3.3), targeting doping fractions from 8 – 26 mol%.

3.2.3.2. Studies of the Local Structure of Zirconia with Low Doping Fractions

Site-selective ($^5D_0 \leftarrow {}^7F_0$) Eu^{3+} TRLFS Studies

Excitation and emission spectra

In the excitation spectra of the low doped phases, where a purely monoclinic phase is present, it can be seen that four species (marked with dotted lines in Figure 33, left) with differing excitation energies are present. Two of the species in the sample with 500 ppm Eu^{3+} doping, one at an excitation maximum of 574.5 nm (excitation peak 1 in Figure 33, left) and one at 582.6 nm (excitation peak 4), show very narrow excitation peaks with FWHMs of around 0.4 and 0.2 nm, respectively. These narrow excitation signals must arise from a rather defined environment of the Eu^{3+} ion. Between these two peaks, two additional, broad excitation signals with maxima at approximately 577 nm (excitation peak 2) and 578.8 nm (excitation peak 3) and half widths of around 2 nm can be seen. The broad excitation bands point towards a more disordered Eu^{3+} environment within the solid structure. At a higher doping percentage of 0.5 mol% Eu^{3+} the intensity of the narrow peak, peak 1 decreases while the opposite can be seen for the second narrow peak, peak 4. In addition, peak 2 decreases in intensity while peak 3 shifts to about 579.3 nm. At a doping level of 2 mol% only the intensity of peak 4 changes drastically, causing this peak to almost disappear while the rest of the excitation spectrum stays rather constant.

Peak 3 appears in all samples with Eu^{3+} doping, and it becomes the sole peak in the excitation spectra of the tetragonal and cubic samples (from 8 mol% doping fraction), as will be discussed in chapter 3.2.3.3. Conclusively, this peak must result from Eu^{3+} incorporation within the ZrO_2 lattice, resulting in stabilization of the crystal structure. Based on this assignment, the similar broad signal of peak 3, present in the monoclinic samples is assigned to incorporation of Eu^{3+} into zirconia as well.

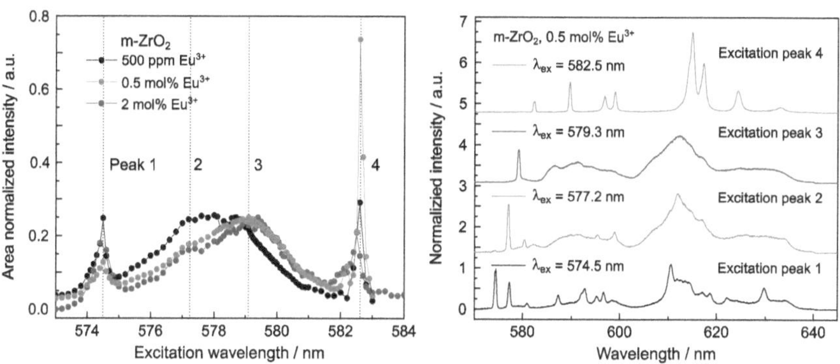

Figure 33: Selected excitation spectra of monoclinic zirconia samples doped with 500 ppm, 0.5, and 2 mol%, with markers for the excitation peak positions (left) and emission spectra excited at the marked positions in the excitation spectrum, plotted with increasing offset (right).

The emission spectra from the monoclinic sample doped with 0.5 mol% Eu^{3+}, collected at all four excitation peak maxima is presented in Figure 33, right. The emission after excitation at 574.5 nm (excitation peak 1) shows numerous very narrow peaks overlapped by broad emission peaks which coincide with the emission signal of the excitation peak 2 and 3. Therefore, and from the fact that multiple peaks can be seen in the $^5D_0 \rightarrow {}^7F_0$ transition, it can be assumed that energy transfer takes place between the defined species of the highest energy (excitation peak 1) to the less defined, broad species of lower energy (excitation peak 2 and 3). The emission spectrum excited at peak 2 shows broad, low symmetry emission spectra, as expected from the broad excitation peak. Here again, multiple species overlap, as can be seen from the $^5D_0 \rightarrow {}^7F_0$ transition, showing two peaks. The narrow excitation peak 4 yields an emission spectrum of clearly resolved, narrow peaks with a low symmetry splitting pattern (i.e. threefold and fivefold for the $^7F_1$ and the $^7F_2$ bands, respectively). The emission spectrum at peak 3, which is assumed to result from the incorporation of Eu^{3+} into zirconia, gives very broad peaks with a low symmetry splitting. The maximum peak intensity is at 612.0 nm. The additional species appearing in the excitation and emission spectra must result from Eu^{3+} species which are not incorporated on the host lattice site of monoclinic zirconia. Especially the species of excitation peak 4 shows a strikingly differing emission behavior.

As described in chapter 5.5.2.2.1, these samples have been prepared by co-precipitation of Zr^{4+} and Eu^{3+} as hydrous oxide and hydroxide, respectively, followed by calcination in air. As a result of this procedure, any Eu^{3+} which is potentially not incorporated into ZrO_2 would remain sorbed to the surface of ZrO_2 in suspension, and, in the calcination transform into Eu_2O_3. Various luminescence spectra can be found in literature for Eu_2O_3 as well as Eu^{3+} doped REE_2O_3. The luminescence behavior of Eu^{3+} in these systems, however, depends on the size of the solid phase as well as its crystal structure.[156] Therefore, to elucidate whether the presence of Eu_2O_3 could cause additional luminescence signal, pure Eu^{3+} has been precipitated as hydroxide which was then calcined in air yielding Eu_2O_3 with a cubic crystal symmetry of the space group $I A\bar{3}$. The excitation spectrum of this pure europium(III) oxide sample and the monoclinic zirconia, where a small amount of Eu^{3+} was added during co-precipitation (0.5 mol%), are plotted together in Figure 34, left.

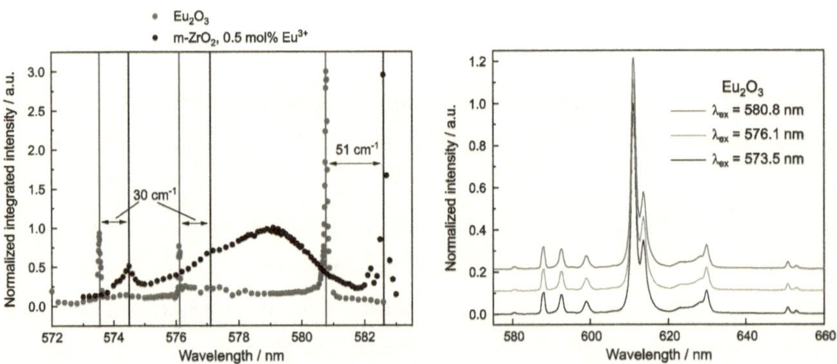

Figure 34: Excitation spectrum of pure Eu₂O₃ (blue) in comparison to a low doped monoclinic zirconia sample (black, left), with lines indicating shifts of the peak positions. Emission spectrum of Eu₂O₃ excited at its three excitation peaks, plotted with increasing offset (right).

Despite many obvious differences in these spectra, the three excitation peaks of the Eu_2O_3 at peak positions of 573.4, 576.0, and 580.6 nm seem to resemble those in the monoclinic zirconia sample as well, however, shifted to 574.4, 577.0, and 582.3 nm. The energy shift (doped ZrO_2 vs. Eu_2O_3) of both peaks at higher energies (and shorter wavelengths) is 30 cm^{-1}, while the peak at lower energies (and higher wavelengths) is shifted by 51 cm^{-1}. The similarities of the species are most clear at the peak position of lowest energy (highest wavelength), where the peak width of both peaks is very similar. The other two peaks of the zirconia sample are partly covered up by the broad excitation peak with a maximum of 579.3 nm.

In the case of the pure Eu_2O_3 two lattice sites are present, one of higher site symmetry (S_6), which can be assigned to the excitation peak at 576.0 nm and one of lower site symmetry (C_2), which can be assigned to the excitation peak at 580.6 nm.[157] The emission at all three excitation peak positions yields the exact same spectrum (Figure 34, right). One of the reasons is, that energy transfer of the higher symmetry state (S_6, excitation peak at 576.0 nm) to the lower symmetry state (C_2, excitation peak at 580.6 nm) takes place, as has been described before.[157] The third excitation peak at a wavelength of 573.4 nm has an unusually high energy for a $^5D_0 \leftarrow {}^7F_0$ transition of Eu^{3+}. Therefore, and since it yields the same emission spectrum as the other two excitation peaks, it is assumed that this excitation peak is caused by phonon coupling effects or a RAMAN transition, likely connected to the S_6 site, as both peaks are shifted by the same energy in comparison to the zirconia sample.

The emission spectra of this monoclinic zirconia sample (0.5 mol% Eu^{3+} doping) are plotted together with the corresponding species of the Eu_2O_3 sample in Figure 35.

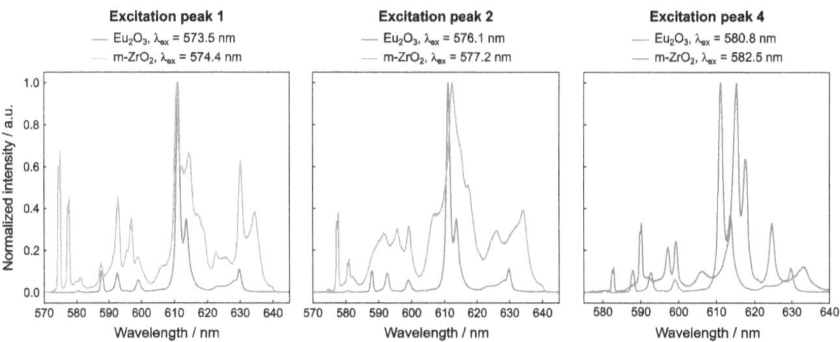

Figure 35: Emission spectra of m-ZrO₂ doped with 0.5 mol% Eu³⁺ and of Eu₂O₃, both excited at their lowest excitation peak position (573.5 nm and 574.4 nm, left), at 577.2 nm and 580.8 nm (middle) and at their highest excitation peak position (580.8 nm and 582.5 nm, right).

The defined peaks after excitation at 574.4 nm (excitation peak 1) correlate with the ones of Eu_2O_3 excited at 573.5 nm, which overlap with the broad emission peaks discussed before (Figure 35, left). When comparing the emission of the zirconia sample at 577.2 nm (excitation peak 2) to the one of Eu_2O_3 excited at 576.1 nm a similar peak pattern, i.e. similar intensity ratios and peak widths, can be seen for the narrow peaks in the zirconia samples spectrum, however, slightly shifted to lower energies (Figure 35, middle). The narrow peaks after excitation at 582.5 nm (excitation peak 4) yield, once again, very comparable intensity ratios and peak widths as Eu_2O_3, but shifted to lower energies (Figure 35, right).

From these observations, it can be concluded that a Eu_2O_3-like structure forms in these samples which is affected by the zirconia environment causing an additional perturbation of the electronic levels of the excited states and/or ground state, subsequently resulting in slightly shifted luminescence spectra. The S_6 site is less affected by the zirconia environment, likely due to its higher symmetry. The C_2 site, however, shows a much stronger shift in the transition leading to the conclusion that this site of lower symmetry is more disturbed by the zirconia environment. Further studies will be necessary to fully describe this Eu_2O_3-like structure. However, it is assumed that it can be best described as Eu_2O_3 nanoclusters in a zirconia matrix. These very small clusters cannot be observed with PXRD because of the lack of long range order.[151,158,159] The formation of dopant oxides, when attempting doping levels higher than the solubility limits in ZrO_2, has been observed before for undersized dopants, such as Fe^{3+} and Ga^{3+}.[98]

As a direct result of this, and in contrast to the EXAFS results presented in chapter 3.3.2.2, it can be concluded that a large quantity of the Eu^{3+} dopant is not structurally incorporated into monoclinic zirconia but that Eu_2O_3-like structures form. Therefore, the excitation of the actual structural incorporated Eu^{3+} into m-ZrO_2 takes place at ~ 579 nm only. The reason why this cannot be observed in the EXAFS is that the crystallographic distance of Y–O in Y_2O_3 is 2.25 Å, and the distance obtained from EXAFS for m-ZrO_2 is between 2.26 Å and 2.28 Å. It could be assumed that the actual distance of Y–O in m-ZrO_2 could be slightly larger, however the average distance is reduced due to the presence of Y_2O_3 nano clusters in the structure. Also, the Y–Y distance in Y_2O_3, with a value of 3.46 Å is very similar to the obtained 3.45 – 3.50 Å. Due to the limited availability of reference data in literature on the monoclinic phase specifically, this hypothesis can, however, not be validated.

To get a better insight into the oxide formation process in zirconia, additional m-ZrO_2 samples with low Eu^{3+} doping (50 – 500 ppm) were prepared. The aim of these low-doped samples was to elucidate, whether or not only very small amounts of dopant can be incorporated into the monoclinic zirconia while the excess dopant forms the oxide phase or if incorporation of the dopant is only achieved when enough dopant is available for partial stabilization of the tetragonal phase. The excitation spectra of these samples are shown in Figure 36 for doping levels between 50 and 500 ppm Eu^{3+} together with the excitation spectra of 0.5 and 2 mol% Eu^{3+}, presented before (s. Figure 33, left).

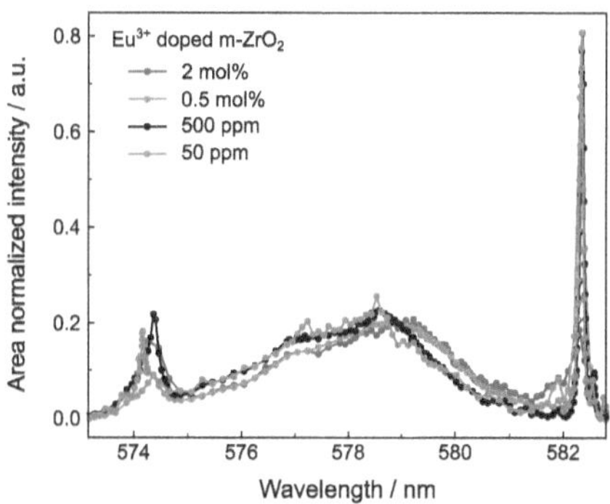

Figure 36: Excitation spectra of m-ZrO_2 doped with various amounts of Eu^{3+} from 50 - 500 ppm (left), and of 0.5 and 2 mol% (right).

From 50 - 500 ppm Eu^{3+} doping, the same four peaks as described above can be observed where very high intensity is obtained for peak 4, i.e. the Eu_2O_3 peak. At higher doping fractions of 0.5 mol% a decrease of the intensity of peak 1, 2, and 4 is accompanied by a small shift of peak 3 to a higher wavelength from 578.8 nm to about 579.3 nm. When further increasing the doping fraction to 2 mol% a drastic decrease of peak 4 is obtained, and peak 3 stays constant. Surprisingly, peak 1 and 2 do not follow the same trend as peak 4, although all three of them are assumed to result from the same species, i.e. nano-particular Eu_2O_3 in the ZrO_2 matrix. As mentioned above, peak 1 is assumed to result from a RAMAN transition, e.g. a phonon-related effect. Due to the presence of multiple phases and sites, multiple such transitions could overlap at the position of peak 1 hindering the observation of a clear trend. A shift and a broadening of this peak 1 could further support this assumption. Peak 2 strongly overlaps with peak 3. Therefore, the increase of peak 3 could cover up a decrease of peak 2.

Lifetimes

To gain further insight into the presence and the amount of oxide clusters in these low-doped samples, the luminescence lifetimes were analyzed in detail. As lifetimes of Eu_2O_3 clusters can be assumed to suffer from severe self-quenching, lifetimes of both non-quenched and quenched Eu^{3+} were analyzed for reference.

The strong self-quenching effect of closely located Eu^{3+} centers can be shown clearly when comparing the lifetime of a pure cubic Eu_2O_3 phase (crystal phase $I\bar{A}3$) to a Eu^{3+} doped Gd_2O_3 phase. The Gd_2O_3 behaves as an analogous structure to the Eu_2O_3 due to the formation of the same crystal phase and comparable ionic radii (95 pm vs. 94 pm at CN 6), where, however, the Eu^{3+} centers are isolated from one another. As can be observed clearly, the pure europium phase shows a strongly reduced lifetime of $\tau_1 = 60 \pm 9$ μs, $\tau_2 = 170 \pm 25$ μs and $\tau_3 = 500 \pm 75$ μs as compared to $\tau_1 = 1800 \pm 270$ μs and $\tau_2 = 3500 \pm 470$ μs in Eu^{3+} doped Gd_2O_3.

The luminescence decay of peak 4 (582.5 nm excitation) lies in between these two extremes, as the quenching probability depends on the amount of Eu^{3+} centers in proximity to the absorbing Eu^{3+}, which is much larger in bulk Eu_2O_3 than in a nano-particular Eu^{3+} cluster surrounded by non-quenching Zr^{4+} cations.

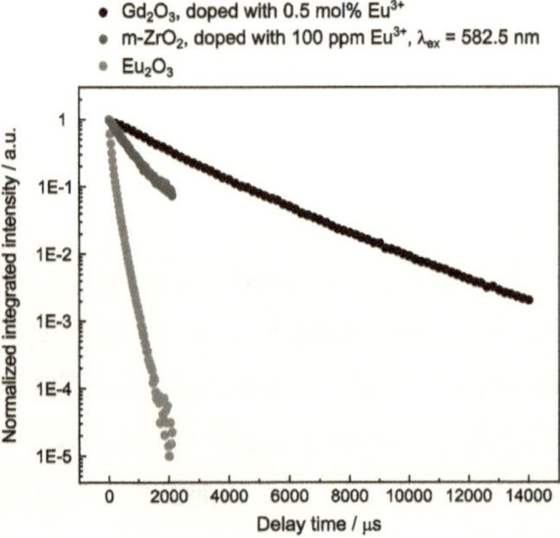

Figure 37: Luminescence lifetime of Gd_2O_3 doped with 0.5 mol% Eu^{3+} (black), m-ZrO_2 doped with 100 ppm Eu^{3+}, excited at peak 4 (λ_{ex} = 582.5 nm, blue) and pure Eu_2O_3 (red).

In Figure 38 the luminescence decay over time is presented for the samples containing 50 ppm to 0.5 mol% Eu^{3+} excited at peak 3 (579.3 nm, squares) and at peak 4 (582.5 nm, dots). It can be observed that the decay, when exciting at peak 3 is slower than when exciting at peak 4. The lifetimes of the former can be fitted using a bi-exponential fit with lifetimes of $\tau_1 = 450 \pm 50$ µs and $\tau_2 = 1950 \pm 250$ µs. Exciting at peak 4 leads to a bi-exponential decay with a lifetime of $\tau_1 = 450 \pm 50$ µs and $\tau_2 = 1250 \pm 200$ µs. The bi-exponential decay of the latter, the faster decay, and the very short lifetime τ_1 of 450 µs is a clear indication for self-quenching taking place for the Eu^{3+} species excited at this wavelength. This observation supports that here, indeed Eu_2O_3 nanoclusters are excited, which will exhibit strong self-quenching, as discussed above. At peak 3, the bi-exponential decay is likely to result from co-excitation and/or energy transfer to the Eu^{3+} species at peak 4, while τ_2 of 1950 µs speaks for isolated Eu^{3+} centers in a zirconia matrix, i.e. Eu^{3+} incorporation.

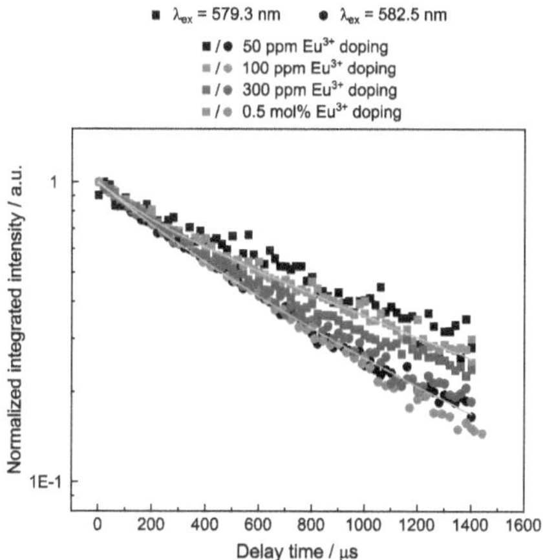

Figure 38: Comparison of luminescence lifetime measurements of m-ZrO_2 doped with 50 ppm to 0.5 mol% Eu^{3+}, excited either at 579.3 nm (squares) or at 582.5 nm (dots). Orange lines are a visualization help only.

Based on the combined results from excitation and emission spectra, as well as the luminescence lifetimes, it appears like both oxide formation and incorporation takes place even at 50 ppm dopant concentrations. Only at higher dopant concentrations where a small amount of phase-transformation from monoclinic to tetragonal zirconia can be observed, a clear decrease of the oxide peak(s) occurs. Due to the presence of several overlapping species in these samples, further studies with Cm^{3+} were conducted as described below.

Cm³⁺ TRLFS Studies

UV-excitation (F ← Z)

To validate the assignment of the incorporation process, some selected Cm^{3+}-doped samples were synthesized. Focus was set on the low dopant range (25, 50, and 100 ppm) to elucidate the oxide formation process in more detail. However, to stabilize the cubic modification an additional sample with high overall doping was also prepared, using Gd^{3+} as the co-dopant. Furthermore, a reference sample of Cm^{3+} doped Gd_2O_3 served as an analogue for Cm_2O_3, which could not be produced in its pure form due to the limited amount and high costs of available ^{248}Cm. However, an analogous behavior of the Gd- and Cm-oxides is expected due the similar cation radii (94 vs. 97 pm for the oxidation state 3+ and six-fold coordination) and the formation of the same crystal phase ($I A \bar{3}$).

The emission spectra of the low-doping concentration row are presented in Figure 39 together with the emission spectrum of the Gd^{3+}/Cm^{3+} co-doped c-ZrO_2 sample (22 mol% Gd^{3+}, 300 ppm Cm^{3+}) and the Cm^{3+} doped Gd_2O_3 sample (50 ppm Cm^{3+}).

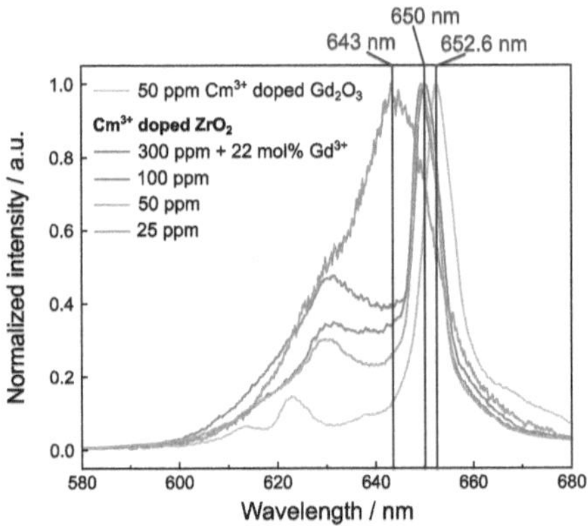

Figure 39: Emission spectra of Cm^{3+} doped m-ZrO_2 at various concentrations (25 – 100 ppm), of Gd^{3+} co-doped ZrO_2 (300 ppm Cm^{3+}, 22 mol% Gd^{3+}) and Cm^{3+} doped Gd_2O_3 (50 ppm Cm^{3+}).

The three m-ZrO_2 samples with low concentrations of Cm^{3+} (25, 50, and 100 ppm) show comparable emission behavior where the main emission peak is found at 650 nm. In addition, a hot–band is resolved at approximately 631 nm for all compositions. It should be noted that 650 nm is an enormous redshift as compared to the Cm^{3+} aquo ion (593.8 nm) which will be discussed in more detail later. Both the hot–band and the main transition peak in these ZrO_2 samples are very comparable to the spectrum obtained for Gd_2O_3 doped with 50 ppm Cm^{3+}, only with a slight additional shift to an emission wavelength of 652.6 nm. In contrast to Gd_2O_3, however, an increasing amount of luminescence signal can be seen in the m–ZrO_2 samples between the hot–band and main transition peaks with increasing Cm^{3+} concentration. A comparison of these low doped samples with the Gd^{3+}/Cm^{3+} co-doped c–ZrO_2 shows that the incorporation in the cubic phase results in a very broad emission peak at 643 nm. This peak coincides with the intensity between the two aforementioned peaks in the low doping row.

Therefore, it is assumed that, in analogy with the Eu^{3+} results, only partial incorporation takes place into the m–ZrO_2 phase, which is a result of the large cation radius mismatch of the host (78 pm) and dopant (97 pm) as well as the small site volume in monoclinic ZrO_2.[31] Further, it is assumed that Cm^{3+} which is not incorporated into the m–ZrO_2 forms a Cm_2O_3 like structure within the solid phase, in analogy to the Eu^{3+} behavior observed above. This hypothesis is based on the observed similarities of the Cm^{3+} emission signal in the m–ZrO_2 samples and in Gd_2O_3. The unit cell volume in Gd_2O_3, is slightly smaller than in Cm_2O_3 due to the differences of the host cation radii (94 vs. 97 pm).[31] This manifests itself as a red–shift of the Cm^{3+} emission signal in the smaller Gd_2O_3 host in comparison to the presumed Cm_2O_3 environment in m–ZrO_2 as seen in Figure 39.

Site-selective excitation (A ← Z)

The site-selective excitation approach clearly shows the presence of multiple phases present in the samples. When exciting at λ_{ex} = 605.0 nm, a very differing intensity ratio can be obtained (Figure 40) as compared to the UV-excitation (s. Figure 39), i.e. a broad emission peak is observed with a maximum of about 638 nm and the narrow peak, presumed to stem from the emission of a Cm_2O_3-like environment is less dominant here.

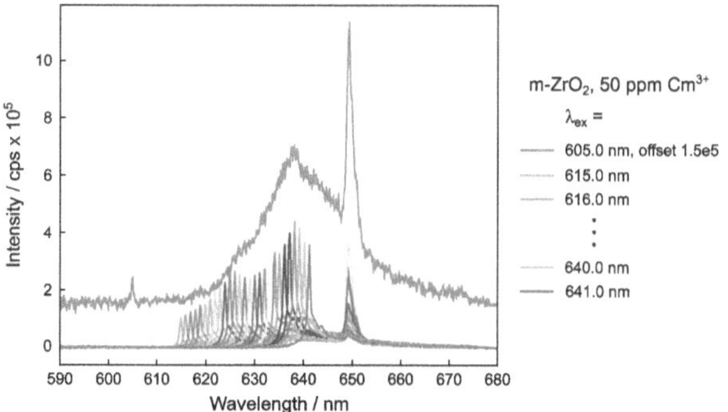

Figure 40: m–ZrO_2 sample, doped with 50 ppm Cm^{3+}, excited at 605.0 nm, plotted with an offset of $1.5 \cdot 10^5$ for better visualization (blue) and at a variety of excitation wavelengths from 615.0 - 641.0 nm, in a step size of 1 nm.

Additionally, a line-narrowing effect can be observed when exciting at wavelengths lying underneath the broad peak at 638 nm, which speaks for a multitude of similar Cm^{3+} environments. This results from a not well-defined environment of the Cm^{3+} site, as seen above for the Eu^{3+} incorporation into zirconia. This supports the hypothesis that Cm^{3+} incorporation into the monoclinic phase takes place. The intensity of the narrow lines caused by the line narrowing effect has an increased intensity at 625 nm, 634 nm, and 638 nm, correlating with two of the hot-band transitions and the peak maximum itself.

When taking a closer look at the sample doped with 100 ppm Cm^{3+}, at certain excitation wavelengths the presence of a second broad species with a maximum of 627.5 nm, in addition to the one at 638 nm can be seen (Figure 41). The same behavior was observed for Eu^{3+} incorporation into m-ZrO_2 (s. chapter 3.0), where two broad excitation peaks (2 and 3) were observed. Therefore, it must be assumed that two differing dopant sites are present in the m-ZrO_2 phase.

A plausible explanation for this would be that a small amount of t-ZrO_2 has formed locally, causing a differing dopant environment. However, the mismatch of the excitation peak positions for Eu^{3+} doping of 578.8 nm vs. 579.3 nm as well as of the emission peaks for Cm^{3+} doping of 638.0 nm vs. 643 nm is in contradiction to this hypothesis.

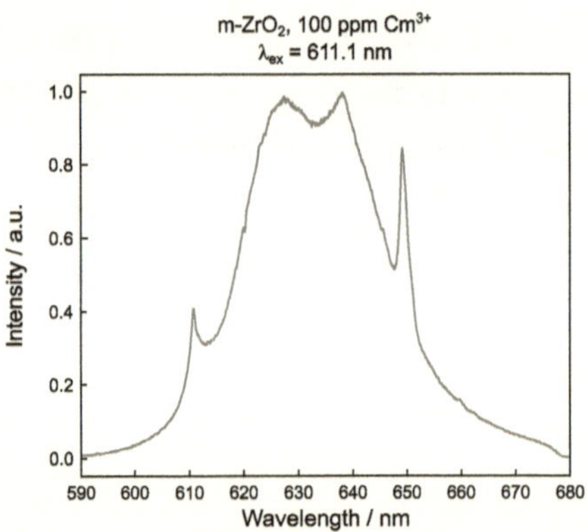

Figure 41: Emission spectrum of m-ZrO_2, doped with 100 ppm Cm^{3+}, excited at 611.1 nm.

The emission of the narrow peak in the Cm^{3+} doped Gd_2O_3 sample as well as in Cm^{3+} doped m-ZrO_2 results from Cm^{3+} in a very defined environment. This can be derived from the very narrow peak shape. Further, close examination shows that the ground-state splitting of Cm^{3+} becomes visible in the spectrum of the Gd_2O_3 sample (Figure 42, left) as well as the ZrO_2 samples (Figure 42, right), which is additional confirmation for a very defined environment.

The Cm-doped ZrO_2 samples could confirm the observations of the Eu^{3+}-doped m-ZrO_2 samples. Additional information, however, could not be obtained even though the overall dopant concentration could be decreased to 25 ppm. Despite of this, the extreme peak-shift obtained for Cm-incorporation in ZrO_2 and the oxide phases is very unusual.

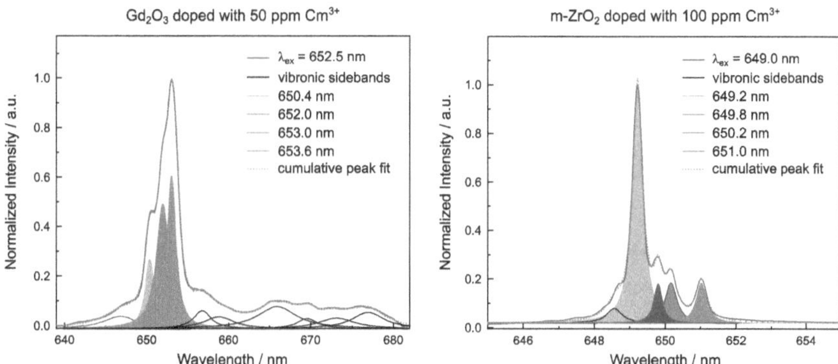

Figure 42: Emission spectrum of Cm^{3+} doped Gd_2O_3 (50 ppm) (left, dark blue) and of m-ZrO_2 doped with 100 ppm Cm^{3+} with fitting of the ground state splitting with four Lorentzian peaks (orange, brown, blue, and red). The black lines indicate additional peaks present in the spectrum, presumably resulting from vibronic sidebands.

In general, a redshift of Cm^{3+} spectra is assumed to result from two factors. One, being a shortening of the bond distance of Cm^{3+} to its coordination sphere and two, the so called nephelauxetic effect, which is the result of a highly covalent bond character.[160] In m-ZrO_2 which is a very small host for the large Cm^{3+} cation, very short bond distances could be assumed. However, our EXAFS studies show that the dopant-oxygen distance in doped m-ZrO_2 is only slightly shortened due to the small site-volume, as compared to the cubic phase. Furthermore, Cm^{3+} doped Gd_2O_3 is not assumed to cause strong bond contraction around the Cm^{3+} due to the similarity of their cation radii. Thus, Zr-bearing solid phases seem to induce an extreme nephelauxetic effect of the Cm-spectra, speaking for a large degree of covalency in these compounds.[161] This is further underlined when comparing reported peak shifts for other Cm^{3+}-containing solid phases. Until now, the largest bathochromic peak shift ever reported for Cm^{3+} was also found for Cm^{3+} incorporation into a Zr(IV)-bearing solid phase, namely pyrochlore $La_2Zr_2O_7$, with a peak maximum at 639 nm.[162] This extreme peak shift was accompanied by a large splitting of the ground-state, of 82 cm^{-1}. In the current study the ground-state splitting in m-ZrO_2 is smaller, i.e. 43 cm^{-1}, which is exceeded by the splitting reported for Cm incorporation in $SrCO_3$ (61 cm^{-1}) and $CaCO_3$ (66 cm^{-1}). In these studies, however, the redshift of the emission peaks was much smaller, i.e. 608.5 nm and 612.7 nm, respectively. Based on these studies, there seems to be no apparent correlation between the magnitude of the crystal field splitting of the ground-state and the energy of the emission transition.

The bonding character of lanthanide oxides is generally not assumed to be very covalent.[163] Therefore, the record-breaking bathochromic shift of 652.6 nm obtained in this study for Cm^{3+} incorporation in Gd_2O_3 and the large ground-state splitting of 76.5 cm^{-1} are not easy to comprehend, and the underlying reasons for these changes in the electronic levels should be explored in more detail. Generally, experimental luminescence spectra for Cm^{3+} incorporation in oxide materials are very scarce. To date, the only reported shift for a pure oxide compound can be found for Cm^{3+} incorporation in ThO_2, with a shift of 626 nm and a ground-state splitting of 36 cm^{-1}.[164] With the available experimental data for ZrO_2, Gd_2O_3, and ThO_2 where the mismatch of the Cm^{3+} vs. host cation size follows the order Zr > Gd > Th, no apparent trend for the peak shifts $\Delta E(Gd) > \Delta E(Zr) > \Delta E(Th)$ can be deduced. Computational studies will be necessary to help understanding this unique behavior.

Lifetimes

The obtained luminescence decay for the Cm^{3+} doping row of 25 ppm to 100 ppm depends on the wavelength region in which the spectra are integrated. When integrating over the range from 635 nm to 645 nm (Figure 43, squares), which corresponds to the range where incorporated Cm^{3+} can be observed, a longer lifetime is obtained as when integrating from 646 nm to 653 nm (Figure 43, dots), where the oxide nanocluster peak is observed for each sample.

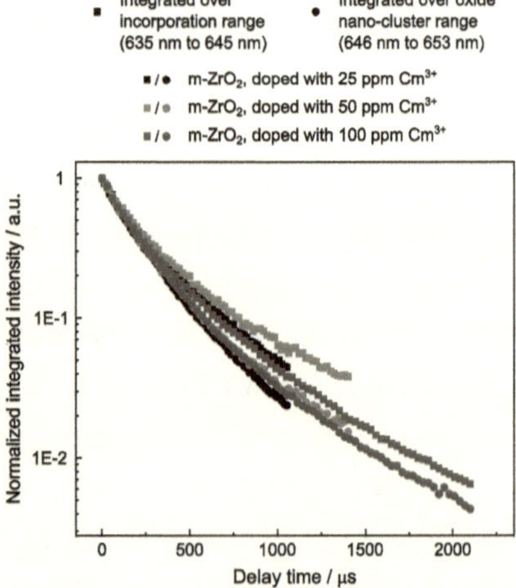

Figure 43: Luminescence lifetimes of m-ZrO_2 doped with 25 ppm to 100 ppm, integrated either over the range from 635 nm to 645 nm (squares), corresponding to the range where signal from the incorporation into m-ZrO_2 is expected or over the range from 646 nm to 653 nm (dots), corresponding to the range of signal from oxide nanoclusters.

All samples and integration ranges can be fitted using a tri-exponential decay curve with τ_1 = 70 ± 20 µs, τ_2 = 240 ± 35 µs and τ_3 = 760 ± 150 µs, however, in differing fractions. Due to the overlapping excitation and emission wavelength of the integration ranges as well as the presence of hot-band transitions of the oxide within the incorporation region, energy transfer cannot be excluded in these samples. Therefore, all samples and integration ranges show lifetimes of both components. The short lifetimes of τ_1 and τ_2 are clear indications for self-quenching effects taking place, as would be expected for Cm_2O_3 nanoclusters. The lifetime of τ_3 equals zero water molecules according to the *Kimura* equation (Eq. 3), and, therefore, speaks for incorporation of Cm^{3+} into ZrO_2. The coefficient of this long lifetime component is larger for the integration over the lower wavelength range (635 nm to 645 nm), i.e. 16%, as when integrating from 646 nm to 653 nm, i.e. 8%. This supports that the emission of the incorporated Cm^{3+} lies in the earlier range and that oxide nanoclusters cause the narrow, stronger redshifted emission peak, as discussed above. This behavior is in analogy to the incorporation of Eu^{3+} into m-ZrO_2, where differing lifetimes are obtained depending on which species is preferentially excited.

The lifetime of the Cm^{3+} decay, integrated over the oxide range (646 nm to 653 nm) exhibits significant self-quenching as compared to the lifetime of Cm^{3+} incorporated into the analogues Gd_2O_3 structure (Figure 44). The preparation of a pure Cm_2O_3 component in analogy to the pure Eu_2O_3 presented above is not possible due to the scarcity of curium. However, it is assumed that an even shorter lifetime would be obtained for such a pure component due to the strong self-quenching effect.

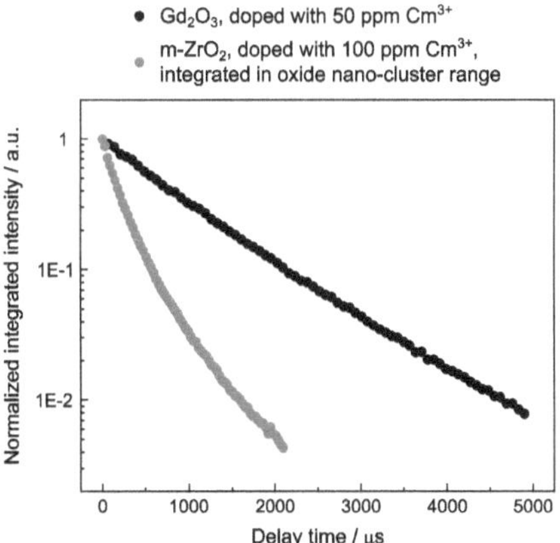

Figure 44: Luminescence lifetime of Gd₂O₃ doped with 50 ppm Cm³⁺ (black) and m-ZrO₂ doped with 100 ppm Cm³⁺, integrated in the range where signal from the oxide nanocluster is expected.

3.2.3.3. Studies of the Local Structure of Highly Doped Zirconia Phases - Site-selective ($^5D_0 \leftarrow {}^7F_0$) Eu^{3+} TRLFS Studies

Excitation spectra

To gain further insight into the incorporation of Eu^{3+} in stabilized zirconia, such as the potential presence of multiple non-equivalent Eu^{3+} environments in the stabilized solid, as it was observed in m-ZrO$_2$, site-selective TRLFS investigations at T < 10 K were performed. Figure 45, left, shows the excitation spectra of the $^5D_0 \leftarrow {}^7F_0$ transition for Eu^{3+} doped ZrO$_2$ in the dopant concentration range where the mainly tetragonal phase transforms to the cubic zirconia phase.

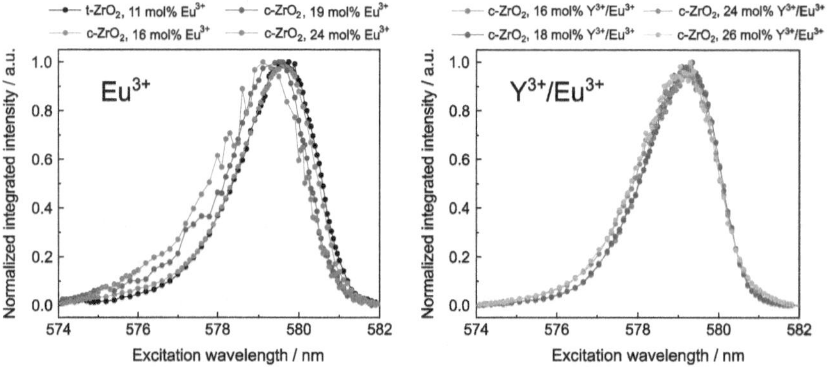

Figure 45: Selected excitation spectra of Eu^{3+}–doped zirconia samples with doping fractions of 11 to 24 mol% (left). Excitation spectra of Y^{3+}/Eu^{3+} co–doped zirconia with doping fractions of 16 to 26 mol% (right).

For the compositions of intermediate doping (11 – 16 mol% Eu^{3+}), one broad excitation peak (FWHM ~ 1.9 nm) at a peak position of approximately 579.3 nm is obtained. This is in contrast to Eu^{3+} incorporation in other crystalline phases, where the incorporation on a crystallographic host lattice site has been shown to yield very narrow excitation peaks (FWHM < 0.1 nm) in the highly ordered, crystalline environments.[32,165] The inhomogeneous peak broadening of the $^5D_0 \leftarrow {}^7F_0$ excitation peak observed here indicates that the environment of Eu^{3+} in the host is not very well–defined but that the excitation peak consist of a multitude of species with slightly differing environments, which results in slightly different excitation energies.[120] This behavior has been observed before for Eu^{3+} doped zirconia,[166,167] as well as other doped phases like tungstates, molybdates,[165] pyrochlores,[162] and several glasses.[168,169]

Typically, an increasing size of the crystal lattice, derived from extracted lattice parameters has been shown to result in a very systematic blue–shift of the Eu–excitation peak due to the weaker ligand field exerted on the Eu^{3+} cation by the coordinating oxygen ligands.[32,165] However, a simultaneous increase of the Eu-O bond length has been shown in these studies, explaining the blue shift of the excitation peak as the Eu-O distance becomes longer and, subsequently, the ligand-field effect weaker. Based on the EXAFS results in the present work, the dopant-O distance was shown to remain practically unchanged over the whole investigated compositional range where the cubic phase predominates (s. Table 7). However, a clear peak shift toward higher energies occurs for higher substitutions, i.e. for 24 and 26 mol% (Figure 45, right).

The shifting of the excitation peak despite of a constant average Eu-O bond distance points towards the presence of multiple distinct dopant environments in the solid structure, i.e. Eu^{3+} sites with differing coordination environments. The presence of multiple Eu^{3+} environments in stabilized ZrO_2 solids has been reported in multiple studies. Montini et al.[170] have studied Eu^{3+}/Ce^{4+} co–doped zirconia phases, and found clear evidence for the presence of three non–equivalent Eu^{3+} environments in the stabilized host structures based on the collected Eu^{3+} excitation spectra. The authors attributed the more red–shifted species with excitation peak positions of 579.3 nm and 580.5 nm to two non–equivalent incorporation species on two differing sites in the solid matrix. A blue shifted component with an excitation peak position around 578.5 nm was shown to arise from Eu^{3+} incorporation into a superficial site. In studies by Yugami et al.[171] and Borik et al.[172,173], three to four different Eu^{3+} environments in Y^{3+}/Eu^{3+} co–doped polycrystalline samples or single crystals of zirconia, respectively, were observed. The authors attributed the specific emission signals to the formation of oxygen vacancies in the crystal structure and the subsequent change in the oxygen coordination number around the Eu^{3+} cation depending on the location (nearest–neighbor (NN) or next–nearest–neighbor (NNN) positions) of the oxygen vacancies.

Thus, in analogy with these studies, the derived excitation profiles were fitted using multiple Gaussian peaks. The best result was achieved using three Gaussian peaks to fit the overall excitation spectra (see Figure 46 for two examples).

The peak positions for the samples from 11 to 24 mol% Eu^{3+} and 16 to 26 mol% Y^{3+} doping were first fitted by varying all parameters. Thereafter, the obtained average values for the peak positions of the present species, namely 578.1 ± 0.5 nm (species 1), 579.0 ± 0.4 nm (species 2), and 579.7 ± 0.5 nm (species 3) were fixed to avoid overparameterization. All samples could be fitted with these peak positions. The results of the fitting of all excitation spectra are summarized in Table 8.

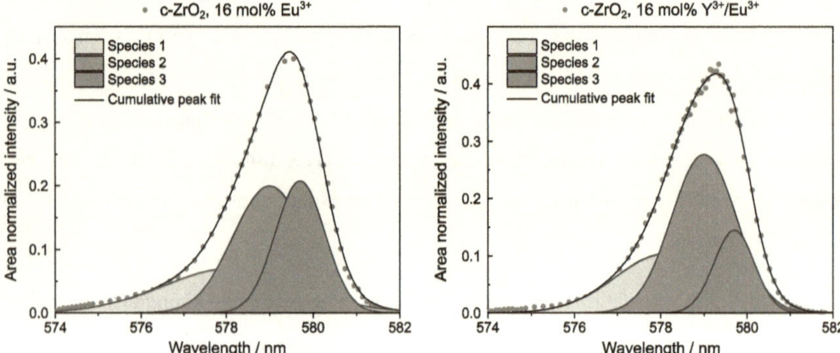

Figure 46: Gaussian fits of excitation spectrum collected for 16 mol% Eu^{3+}–doped (left) and Y^{3+}/Eu^{3+} co–doped c–ZrO_2 (right).

Table 8: Summary of the results from the Gaussian fitting of the excitation spectra from TRLFS. Fixed values are marked with an asterisk (*).

M^{3+} doping / mol%	Species 1			Species 2			Species 3		
	λ_c / nm	%	FWHM / nm	λ_c / nm	%	FWHM / nm	λ_c / nm	%	FWHM / nm
Eu^{3+}									
11	578.1*	22	2.47	579.0*	37	1.84	579.7*	41	1.31
13	578.1*	24	2.91	579.0*	46	1.75	579.7*	30	1.20
16	578.1*	28	3.19	579.0*	42	1.69	579.7*	30	1.15
19	578.1*	48	3.00	579.0*	35	1.26	579.7*	17	0.94
24	578.1*	58	2.83	579.0*	30	1.31	579.7*	12	1.00
Y^{3+}/Eu^{3+}									
16	578.1*	30	2.31	579.0*	52	1.49	579.7*	18	0.95
18	578.1*	31	2.29	579.0*	51	1.45	579.7*	18	0.99
24	578.1*	38	2.65	579.0*	51	1.57	579.7*	11	0.92
26	578.1*	37	2.67	579.0*	55	1.58	579.7*	8	0.88

A systematic decrease of the FWHM can be observed from species 1 to species 3. In both the Eu^{3+}–doped and Y^{3+}/Eu^{3+} co–doped samples, species 1 increases significantly for samples of high doping where a clear shift of the excitation peak was observed, while species 2 varies slightly in the range of 30 to 45% for Eu^{3+} doped samples and 51 to 55% for Y^{3+}/Eu^{3+}

co–doped samples. Species 3 decreases in both cases with increasing doping fraction to about 10% for the highest doping fractions used here.

The obtained peak position for species 1 (578.1 nm) agrees rather well with the peak position of 578.5 nm reported by Montini et al.[170] for Eu^{3+} incorporation at the zirconia surface. In the zirconia single crystals studied by Borik et al.[172,173] such a blue shifted $^7F_0$ peak position was not observed. However, given the much smaller surface area of a single crystal compared to nano–particular powder, no significant contribution of a surface associated Eu^{3+} species to the luminescence is expected. The increasing formation of species 1 with increasing doping percentage follows the trend of decreasing crystallite size seen in the PXRD bulk studies (s. Table 5). Thus, it is likely that the Eu^{3+} dopant is located on a near-surface site, which will increase in abundance as the crystallite size decreases. This is visualized in Figure 47 for the Eu^{3+} doping row (left) and the Y^{3+}/Eu^{3+} co-doping row (right). This hypothesis is further supported by the larger abundance of species 1 in the Eu^{3+} doped ZrO_2 solid phases, which are smaller in crystallite size, in comparison to the Y^{3+}/Eu^{3+} co–doped ones, which yield larger crystallites. Due to lattice relaxations at a superficial site, a higher degree of freedom and therefore, a greater variety of coordination spheres are possible, resulting in the rather large FWHM in comparison to the two other species.

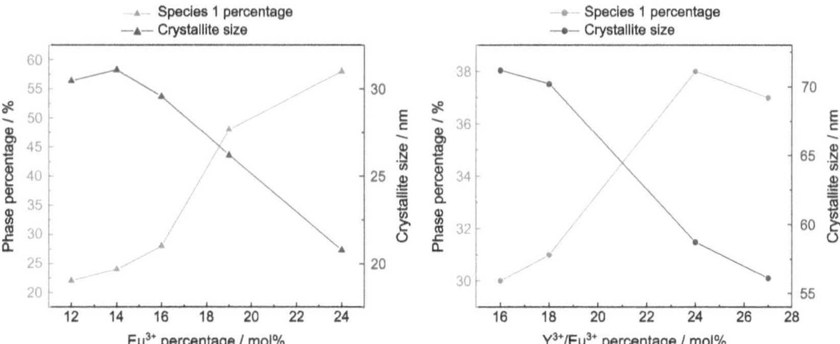

Figure 47: Comparison of the phase percentage of species 1 in the site–selective TRLFS excitation spectra, determined by Gaussian peak fitting with the crystallite size derived from PXRD for Eu^{3+} doped (left) and Y^{3+}/Eu^{3+} co–doped zirconia samples (right). The green data points in both figures refer to the species fraction scale on the left y-axis and the crystallite size, in blue, corresponds to the scale on the right y-axis.

The excitation peak positions of species 2 (579.0 nm) and species 3 (579.7 nm) are very similar to those identified by Borik et al.[173] at 578.9 nm and 579.3 nm. Excitation spectra with similar excitation peak positions can also be deduced from the data in Yugami et al.[171], however, exact peak positions for the extracted species were not given. In accordance with the observations in this study, Yugami et al.[171] reported a decreasing fraction of the excitation peak at ~ 579.8 nm with increasing doping fractions and the conclusive attribution to Eu^{3+} coordinated to eight oxygen atoms in the crystal structure.

In order to visualize the trends of the various Eu^{3+}–species derived in the present study, the mole percentages of all Eu^{3+} species and the number of oxygen vacancies have been plotted as a function of the Eu^{3+}–doping in Figure 48, left and $Eu^{3+} + Y^{3+}$ doping fractions in Figure 48, right.

A line for the distribution of oxygen vacancies around the M^{3+} ion (orange line), assuming a random (non–preferential) distribution of the vacancies in the crystal lattice, was included in the figures. More details on these plots can be found in relation to Figure A1. In both the pure Eu^{3+} doping row as well as the Y^{3+}/Eu^{3+} co–doping row, species 2 can be seen to correlate well with the overall number of oxygen vacancies (black dashed line). Especially in the Y^{3+}/Eu^{3+} co-doping samples (Figure 48, right) the amount of this species 2 (green, filled squares) and the number of oxygen vacancies are almost identical. In case of solely Eu^{3+} doped ZrO_2 (Figure 48, left) an increase of the amount of species 2 from 11 to 16 mol% doping can be seen (green, filled circles), after which a plateau is obtained. In this plateau region, a strong increase of the surface associated species (species 1) occurs, which may explain the lower amount of species 2 in these Eu^{3+}–doped samples in comparison to the Y^{3+}/Eu^{3+} co–doped ones. In general, however, this species clearly follows the trend of the increasing amount of oxygen vacancies in the host lattice with increasing M^{3+} doping, which implies that this Eu^{3+} environment is associated with an oxygen vacancy in the first coordination sphere, i.e. in NN position.

Interestingly, for dopant percentages of 11 – 16 mol% (Eu^{3+} doped samples) and of 13 – 26 mol% (Y^{3+}/Eu^{3+} co–doped samples) the amount of Eu^{3+} species with the oxygen vacancy in NN position (species 2) is larger than the statistical distribution of vacancies between Zr and Eu. This implies that oxygen vacancies have a slight tendency for the dopant Eu^{3+} in these samples. Even though both Yugami et al.[171] and Borik et al.[173] reported on the relative amounts of Eu^{3+} species with one or two vacancies in NN or NNN positions in their ZrO_2 host phases, this study is the first to report on the preferential location of oxygen vacancies for Eu^{3+}–doped ZrO_2 samples. There are experimental[174] as well as computational[175] studies supporting the results obtained in the present study, where a preferential location of oxygen vacancies in the crystal lattice was found in the direct coordination of other oversized trivalent dopants rather than Zr^{4+}. These studies, however, are outnumbered by both experimental results and computational evidence for the preferential location of oxygen vacancies around Zr^{4+}. This vacancy formation around the Zr host is explained to promote the stabilization of the tetragonal and cubic zirconia phases at ambient conditions as the reduction of the Zr–O coordination from eight to seven reduces the stress around the rather small host cation. Therefore, there is reason to believe, that the sample synthesis procedure in terms of calcination time and temperature has a strong influence on the location of oxygen vacancies in the crystal lattice.

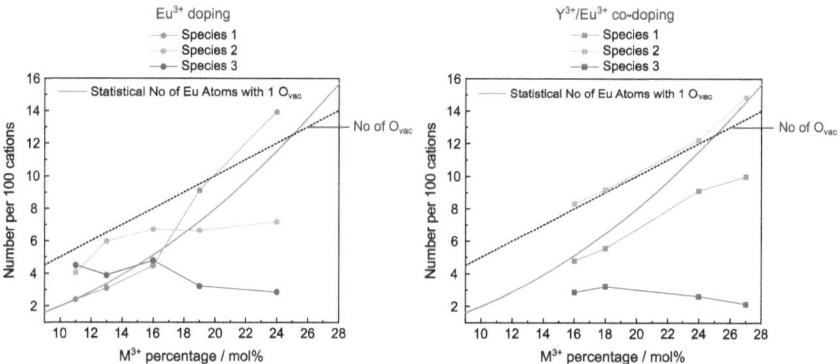

Figure 48: Number of oxygen vacancies in dependence of the Eu^{3+} doping fraction (black, dashed line) and number of cations with an oxygen vacancy in the first coordination sphere of Eu (orange line), assuming a non–preferential, statistical distribution. Number of species 1 (purple), species 2 (green), and species 3 (blue) Eu cations in the pure Eu^{3+} doping row (dots, left) as well as the Y^{3+}/Eu^{3+} co–doped row (squares, right) per 100 cations.

These samples are not in thermodynamic equilibrium due to the very slow cation diffusion in ZrO_2 even at 1000 °C, which, however, is a rather typical calcination temperature in the synthesis of ZrO_2.[145,146,176,177] Although the anion mobility in stabilized zirconia is higher, the same assumption can be applied for the oxygen distribution in the crystal lattice. When the dopant is introduced into the zirconia solid phase, vacancy formation occurs next to the dopant due to the charge mismatch ($M^{3+} \leftrightarrow M^{4+}$).[175] For short synthesis times, especially when combined with rather moderate temperatures, oxygen migration through the lattice may not reach a steady state which subsequently results in an unexpected distribution of oxygen vacancies in the crystal lattice. Thus, it is assumed that an increase of the calcination time and temperature would drive the vacancies toward the host cations.

Emission spectra

High–resolution emission spectra were recorded at varying excitation wavelengths for Eu^{3+}–doped and Y^{3+}/Eu^{3+} co–doped zirconia solid solutions. In Figure 49 emission spectra collected at an excitation wavelength of 579.3 nm (overall excitation peak maximum, see Figure 45) are presented for selected M^{3+} doped ZrO_2 compositions.

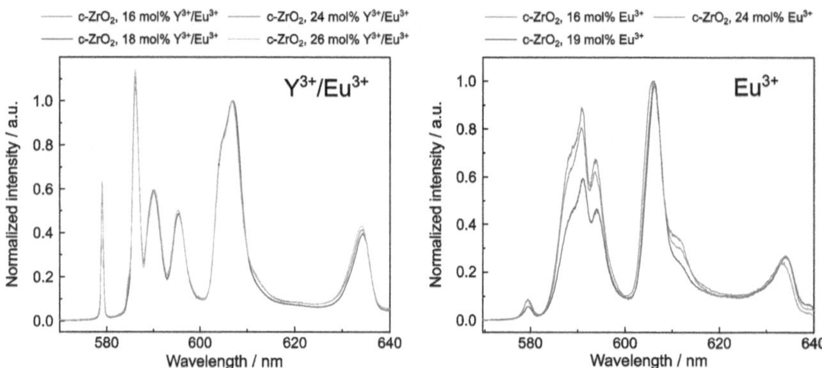

Figure 49: Emission spectra of Y^{3+}/Eu^{3+} co–doped zirconia doped with 16 – 26 mol% (left) and emission spectra of Eu^{3+} doped zirconia with doping percentages from 16–24 mol% (right), λ_{ex} = 579.3 nm, normalized to the peak maximum of the $^5D_0 \rightarrow {}^7F_2$ transition due to the presence of defect luminescence especially prominent in the Y^{3+}/Eu^{3+} co–doped samples.

A clear three–fold splitting of the $^5D_0 \rightarrow {}^7F_1$ transition can be observed for the Y^{3+}/Eu^{3+} co–doped samples (Figure 49, left). However, it should be noted that the first peak of the $^5D_0 \rightarrow {}^7F_1$ transition is overlapped by defect luminescence resulting from the M^{3+} incorporation. This defect luminescence was observed before in similar systems and will be described in detail in the appendix.[162,178] The equivalent Eu^{3+} doped samples yield a rather broad $^5D_0 \rightarrow {}^7F_1$ peak, where the threefold splitting is not as well resolved (Figure 49, right). It is, however, clear that the expected splitting pattern for Eu^{3+} incorporation in a cubic environment, i.e. 1–fold and 2–fold splitting of the $^7F_1$ and $^7F_2$ bands, respectively, is not obtained in any of the examined solid phases. Instead the degeneracy is fully lifted by the surrounding ligand field, corresponding to Eu^{3+} incorporation in a low symmetry environment. The $^7F_2$–band shows a very peculiar splitting pattern. For low symmetry systems, a full 5–fold splitting of this band is expected. Instead only three resolved signals can be deduced from the spectra. In addition, the $^5D_0 \rightarrow {}^7F_2$ transition is different in the Y^{3+}/Eu^{3+} co–doped samples than in the purely Eu^{3+}–doped ones. In the Y–containing samples, the main peak is shifted from 606 nm to 607 nm and it shows a shoulder on its blue side, at 605 nm. The shoulder on the red–side of the main peak of the Eu^{3+} doped samples at

611 nm can, however, not be observed in the Y^{3+}/Eu^{3+} co–doped samples. The origin for this difference must be related to the presence of different Eu^{3+} environments in the solid phases.

In purely, high doped Eu–doped samples, excitation energy transfer between adjacent Eu^{3+} species will take place causing an overall quenched luminescence signal. The excitation energy transfer hampers any distinction between non–equivalent species since the emission spectra are almost identical, independent of the incident excitation wavelength. Further insight into the energy transfer effect is given in Figure 50, where a comparison of the emission patterns of a zirconia sample doped with Eu^{3+}, and of a Y^{3+}/Eu^{3+} co–doped sample, both excited at 579.3 nm is plotted. It can be observed that the emission peak positions match very well. Therefore, it is assumed that the emission from the purely Eu^{3+} doped samples, which is the same for all the excitation wavelengths used in this study, due to energy transfer processes, results from the same Eu^{3+} site as the emission of the Y^{3+} co–doped sample, excited at 579.3 nm.

To avoid energy transfer effects the following discussion of the luminescence emission behavior is focused on the Y^{3+}/Eu^{3+} co–doped zirconia solid–solutions. Due to the low overall Eu^{3+} doping in these samples (0.5 mol%), full isolation of Eu–centers can be assumed and no excitation energy transfer between non–equivalent Eu^{3+} species takes place.

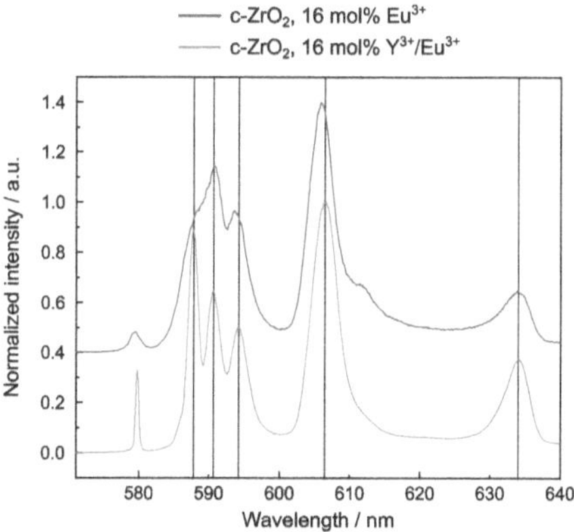

Figure 50: Comparison of the emission spectra of Eu^{3+} doped zirconia and of Y^{3+}/Eu^{3+} co–doped zirconia, plotted with an offset, excited at λ_{ex} = 579.3 nm. Black lines are a visualization help only.

Figure 51 shows emission spectra of a Y^{3+}/Eu^{3+} co–doped sample collected at three different excitation wavelengths. Emission spectra recorded using an excitation wavelength of 577.2 nm, are clearly different from the emission recorded at 579.3 nm (Figure 51, left). In the $^5D_0 \rightarrow {}^7F_1$ transition all three peaks are broadened in comparison to the ones from excitation at 579.3 nm. Especially the second and third peaks seem to be caused by an overlap of two unresolved peaks each. The main peak of the $^5D_0 \rightarrow {}^7F_2$ emission at 579.3 nm has a peak maximum at 607.0 nm with the aforementioned shoulder at about 605 nm. When exciting at 577.2 nm the transition is split into two peaks, one at 603.1 nm and one at 609.2 nm while the latter one is also broadened on the blue side, indicating the presence of another species. Due to overlapping excitation energies of the species, emission peaks belonging to the Eu^{3+} species excited at 579.3 nm can also be deduced in the spectrum recorded at $\lambda_{ex} = 577.2$ nm. However, this rather large peak splitting in the $^7F_1$ as well as the $^7F_2$ band indicates a rather disordered Eu^{3+}–environment.

When comparing the emission after excitation at 579.3 nm and 580.0 nm rather similar spectra are obtained (Figure 51, right) which differ from the one obtained at $\lambda_{ex} = 577.2$ nm, regarding the peak positions as well as the considerably smaller peak splitting within the $^7F_1$ and $^7F_2$ bands. Interestingly, the splitting of the $^7F_1$ band is smaller for the Eu^{3+} environment excited at $\lambda_{ex} = 580.0$ nm than for $\lambda_{ex} = 579.3$ nm, resulting in slightly different $^7F_1$ band positions. For $\lambda_{ex} = 580.0$ nm the main peak is located at 606.4 nm, compared to at 607.0 nm for 579.3 nm excitation and it is slightly narrower, pointing toward a more ordered Eu^{3+}–environment. The overall magnitude of the crystal field splitting of the $^7F_1$ band decreases in the same order pointing toward a lowering of the crystal–field perturbation.

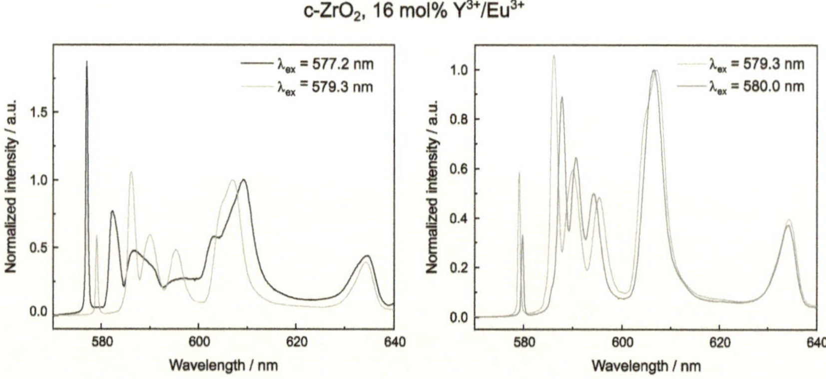

Figure 51: Comparison of emission spectra of Y^{3+}/Eu^{3+} co–doped ZrO_2 samples (16 mol% Y^{3+}/Eu^{3+}) excited at 577.2 nm vs. 579.3 nm (left) and excited at 579.3 nm vs. 580.0 nm (right).

This is in line with the recorded excitation spectra at excitation wavelengths of 577.2 nm (predominantly exciting species 1), 579.3 nm (predominantly exciting species 2), and 580.0 nm (predominantly exciting species 3), where narrower excitation peaks were observed from species 1 to species 3. In addition, the overall magnitude of the crystal field splitting of the $^7F_1$ band decreases in the same order pointing toward a lowering of the crystal–field perturbation. Thus, there seems to be a systematic increase of the local order in these three Eu^{3+} environments (species 1 → species 3) which is in line with the assignment based on the excitation spectra.

Luminescence Lifetimes

The lifetimes of the Eu^{3+} doped zirconia phases have been recorded for additional insight into the species present in the system. It can be observed that the luminescence lifetimes remain fairly constant after excitation at 579.3 nm with increasing Eu^{3+} doping until about 11 mol% (Figure 52). A bi–exponential fit yields two lifetimes τ_1 = 2500 µs and τ_2 = 5000 µs. At higher doping, a clear decrease of the lifetimes is observed. This phenomenon can be assigned to the self–quenching of Eu^{3+}. Therefore, the fitting of the lifetimes at high doping does not provide further information, and, again, the discussion will be focused on the Y^{3+}/Eu^{3+} co–doped samples.

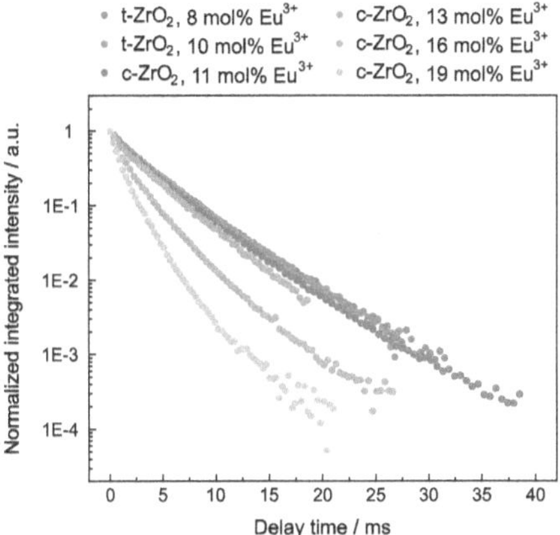

Figure 52: Luminescence lifetimes of Eu^{3+} doped zirconia with doping percentages from 8 – 19 mol% (λ_{ex} = 579.3 nm).

In contrast to the samples doped with Eu^{3+} only, the lifetimes of the Y^{3+}/Eu^{3+} co–doped samples all yield very similar results and no self–quenching effect can be observed. All samples in the row of 16 – 26 mol% Y^{3+} doping and at all three excitation wavelengths (577.2 nm, 579.3 nm, and 580.0 nm) could be fitted by using three different lifetimes, i.e. τ_1 = 1030 ± 310 µs, τ_2 = 2950 ± 250 µs, and τ_3 = 6560 ± 360 µs. Examples of the decay curves and fits are presented in Figure 53. According to the *Horrocks* equation (Eq. (2)), the first lifetime τ_1 yields a coordination of 0.4 H_2O molecules on average while the two longer lifetimes correlate with no water in the coordination sphere.

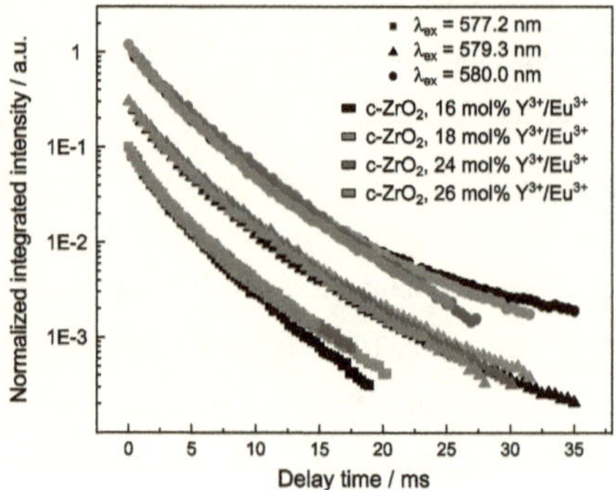

Figure 53: Lifetimes of Y^{3+}/Eu^{3+} co–doped ZrO_2 with doping percentages from 16 mol% to 26 mol% excited at λ_{ex} = 577.2 nm (squares), at λ_{ex} = 579.3 nm (triangles), and at λ_{ex} = 580.0 nm (dots) which are presented with an offset as visualization help. The black lines are examples for the decay curves resulting from fitting the data points.

To assign the lifetimes to the individual species, an excitation spectrum of Y^{3+}/Eu^{3+} co–doped ZrO_2 with an overall dopant concentration of 18 mol% was measured after a delay time of 8 ms and compared to the excitation spectrum recorded 1 µs after the laser pulse (Figure 54). Based on the recorded lifetimes of τ_1 = 1030 ± 310, τ_2 = 2950 ± 250, and τ_3 = 6560 ± 360 µs, the excitation spectrum after a delay of 8 ms should show no presence of the species with the shortest lifetime and approximately 7% and 30% of the species with lifetimes of 2950 µs and 6560 µs, respectively. However, as evident from Figure 54, only a shoulder on the blue side, corresponding to species 1, is completely absent from the delayed excitation spectrum while the main peak has remained almost unchanged. Based on this, the short lifetime of 1030 ± 310 µs of species 1, which excitation peak maximum was observed at 578.1 nm, can indeed be assigned to the incorporation of Eu^{3+} into a near-surface site. This lifetime is rather short and the resulting number of hydration water molecules of 0.4 is rather large when assuming a fully incorporated Eu^{3+} ion. Therefore, the presumed surface layer incorporation of species 1 is supported by the lifetime measurements, as the presence of water molecules on or within the surface layers is likely.

Species 2 and species 3 on the other hand seem to have very similar lifetimes based on the very constant excitation peak shape independent of delay. This is likely to be related to energy transfer mechanisms either from Eu^{3+} to Eu^{3+} centers or, more likely, between Eu^{3+} centers and defect electrons present in the lattice, which hampers the assignment of individual Eu^{3+} environments in the crystal structure based on their luminescence decay.

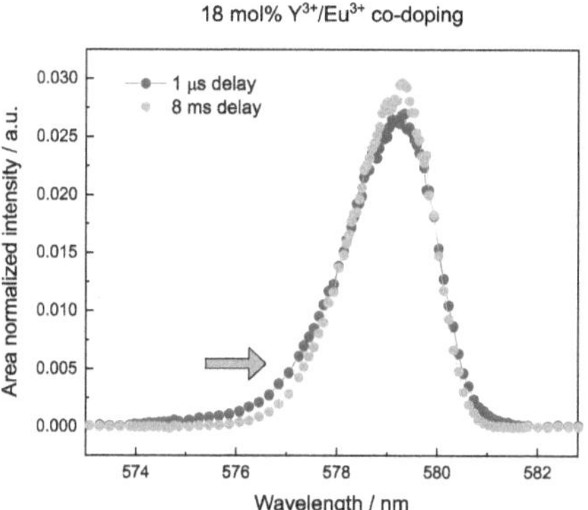

Figure 54: Comparison of the excitation spectrum of a Y^{3+}/Eu^{3+} co-doped zirconia sample (18 mol% doping) measured with a delay of 1 µs and measured with 8 ms delay after the laser pulse. Orange arrow points out main difference in spectra.

Despite not being able to assign the luminescence lifetimes to the three proposed Eu^{3+} environments in the ZrO_2 solids, it is clear that the luminescence spectroscopic data cannot be explained with the presence of one Eu^{3+} environment only. This is in contrast with the results obtained in the EXAFS investigations as there, both the Zr and Y environments stay constant over a large doping range. As EXAFS yields information on the average Zr or Y environment in the sample, the distinction of non–equivalent species in complex samples such as the investigated M^{3+} doped ZrO_2 ones, is hardly possible. In the sample rows studied here, multiple structural changes take place simultaneously. With increasing doping, oxygen vacancies are formed in the lattice, decreasing the overall M–O coordination number. However, at the same time the crystallite size decreases with increasing doping leading to a higher abundance of superficial sites. The changes in the fractions of the three species observed with TRLFS are rather small, further hindering the observation of a trend in the EXAFS fits. Especially the structural difference between species 2 and 3 is vastly defined by a change of the M–O coordination number. However, the determination of the coordination number with EXAFS inhibits a rather large error and the coordination number in ZrO_2 does not seem to be captured properly by EXAFS fitting. This can be seen when attempts are made to fit the coordination numbers (s. Table A1, Table A2, and Table A3), as well as in literature, where the coordination number of doped ZrO_2 is typically fixed to a certain value.[98]

The three different species observed in these studies consist of a sub-set of a large multitude of slightly differing coordination environments themselves resulting from the various possible atomic constellations. This becomes apparent when considering that in a 3 x 3 x 3 supercell of ZrO_2 doped with 16 mol% Y^{3+} a number of $9.3 \cdot 10^{23}$ possible theoretical constellations of Zr^{4+} and Y^{3+} cations, O^{2-} anions and oxygen vacancies exists. This variety of similar species can lead to destructive interference of the EXAFS signal of the individual Eu^{3+} environments, causing a cancellation of their signal. This could affect the accessibility of the differences in this sample row with EXAFS, which can be observed with TRLFS. The large number of oxygen atoms in this shell could make the self–quenching of parts of the signal rather likely.

An alternative explanation for the observed differences in the luminescence behavior of the individual species, but not in the structural studies with EXAFS could be that all three species do in fact have very similar abs–O or abs–Zr distances and thus solely the change of the coordination number and the local symmetry are responsible for the differences observed in the luminescence spectroscopic investigations.

3.2.4. Low Temperature Synthesis Routes of Zirconia

To study the formation and properties of doped zirconia phases at conditions expected to be realistic in a HLW repository, the last step in the synthesis of doped zirconia phases, i.e. calcination was substituted by a low temperature synthesis route. These studies were performed during a research stay at the University of Manchester.

3.2.4.1. Impact of Dopants on Crystallization Rates

Pre-experiments have shown that at 200 °C hydrothermal synthesis conditions, a fully crystalline product is obtained already after 8 h (Figure 55).

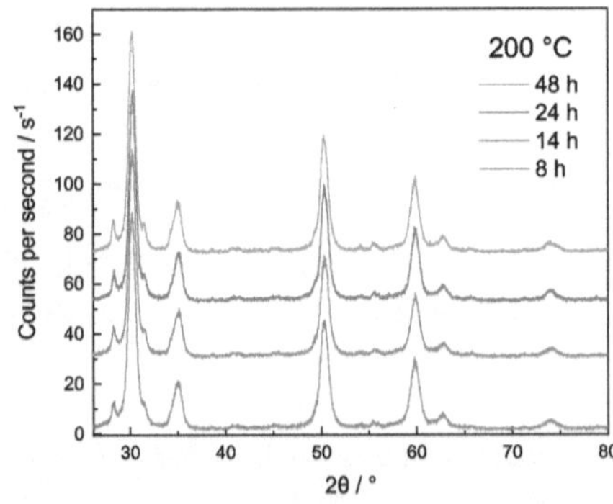

Figure 55: Diffraction pattern of four Eu^{3+} doped (8 mol%) ZrO_2 samples synthesized via low temperature route at 200 °C for 8 – 48 h, measured with Cu K_α radiation. An offset was applied to the individual patterns.

Furthermore, the diffraction pattern does not change when using a hydrothermal treatment time of 8 or 48 h. Therefore, the conditions were changed to 80 °C synthesis temperature of the suspension to enable studying the crystallization mechanism.

Preliminary results on the crystallization rates using no dopant vs. 8 mol% Eu^{3+} doping are presented in Figure 56. Due to the limited amount of data points obtained here, conclusions are rather tentative. However, it seems like the introduction of the Eu^{3+} dopant into the zirconia lattice is reducing the crystallization rates in comparison to pristine zirconia, presumably due to the large mismatch of cation radii (107 vs. 78 pm), hindering the incorporation into the lattice. To validate this, a separate doping row, using Th^{4+} was prepared. Th^{4+} is comparable to Eu^{3+} in terms of the cation radius (105 vs. 107 pm), however, the tetravalent cation does not cause the formation of oxygen vacancies, which reduces the lattice flexibility and which can further be expected to reduce the crystallization rate even more. As can be seen in Figure 56, a strong decrease of the crystallization rate is observed for Th^{4+}, supporting the stated hypothesis. An attempt to obtain more robust data by obtaining more data points to these curves has failed due to unknown issues with the reproducibility of these experiments. To ensure reproducibility, low-temperature treatment with *in situ* PXRD would be necessary, to avoid the artefact introduction when using multiple parallel samples.

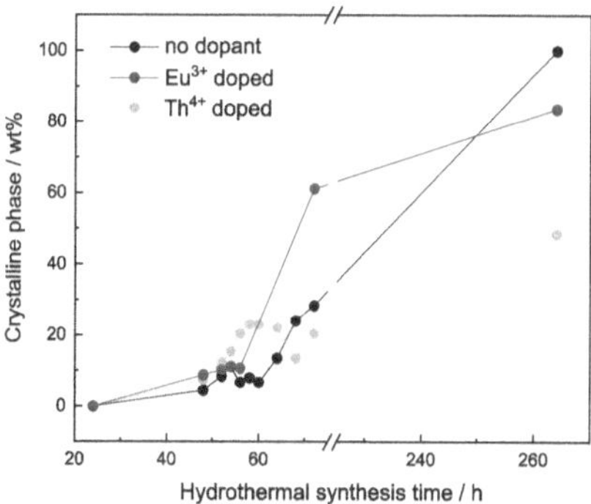

Figure 56: Crystallization rate of hydrothermally synthesized zirconia in dependence of the dopant, i.e. no dopant (black dots), 8 mol% Eu^{3+} (blue dots), and Th^{4+} (orange dots).

3.2.4.2. Impact of Synthesis Time on Crystal Phase

To elucidate, whether the synthesis time affects the crystalline phase of the product, the samples from chapter 3.2.4.1 were studied with PXRD and analyzed using the *Rietveld* refinement.

When studying the crystal phase of a non-doped zirconia sample in dependence of the low temperature synthesis time, a tetragonal phase is obtained after 42 h where a crystalline phase has formed. With increasing synthesis time about half of the crystal phase stays tetragonal while the other half is monoclinic. Only at treatment times longer than 60 h the monoclinic phase starts to dominate. Regarding that the monoclinic phase is the thermodynamically stable phase under these conditions, i.e. no dopant and atmospheric conditions, this is surprising.

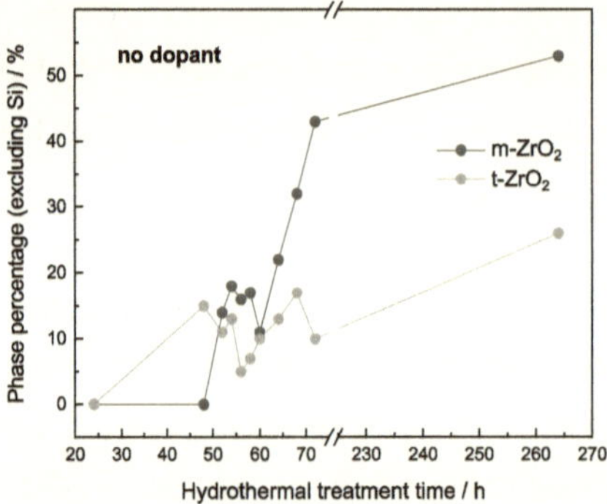

Figure 57: Crystal phase of crystalline part of zirconia in dependence on the synthesis time from 24 to 264 h.

However, it was observed before, that for particles of very small crystallite size, the tetragonal and at even smaller particle size the cubic phase turns stable without the addition of any foreign stabilizers.[95,97] When longer synthesis times are being used, the crystallite size increases causing the formation of the monoclinic phase at a critical size.

While the exact critical sizes for the phase transformation has not been agreed upon, it is generally assumed that the tetragonal to monoclinic transition can be found at a particle size from about 30 – 60 nm.[95] When calculating the crystallite sizes in the critical synthesis time region, it is observed that the m-ZrO$_2$ crystallites are consistently larger than the t-ZrO$_2$ ones. Furthermore, an increasing crystallite size in the monoclinic phase can be seen with increasing time while the tetragonal phase stays rather constant (Table 9).

Table 9: Crystallite size derived by the *Scherrer* equation for the individual phases (m-/t-ZrO$_2$) in dependence of the synthesis time.

Phase\ Synthesis time	54 h	60 h	64 h	72 h
m-ZrO$_2$	16 nm	17 nm	19 nm	20 nm
t-ZrO$_2$	10 nm	8 nm	10 nm	11 nm

This would suggest that under these conditions, the critical size for the phase transition from t- to m-ZrO_2 lies at approximately 10 nm.

3.2.4.3. Eu-concentration Row

To study, whether differing crystal phases are formed in these low-temperature synthesis conditions as compared to the calcination samples studied in chapter 3.3.2.1 - 3.3.2.3, Eu-doped samples with dopant concentrations ranging from 2 to 19 mol% were synthesized.

Therefore, a Eu^{3+} concentration row was prepared using a 16 h heating period to ensure full crystallization. In Figure 58, a comparison of the crystal phase fractions from the calcination route (Figure 58, left) to this hydrothermal route (Figure 58, right) is presented.

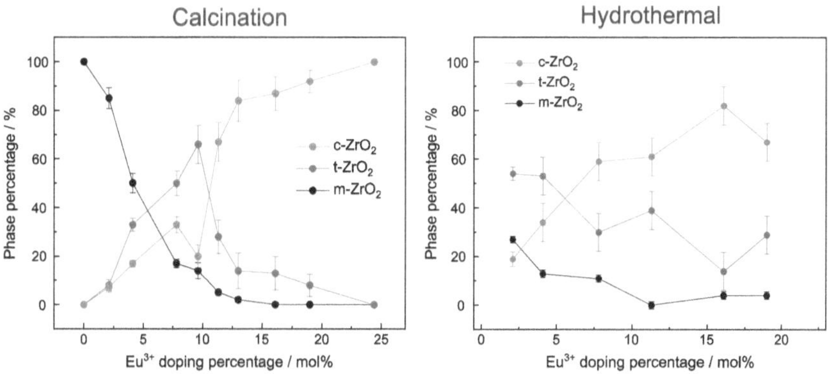

Figure 58: Comparison of phase distribution of samples prepared via calcination (1000 °C, 5 h, left) and hydrothermal treatment (200 °C, 16 h, right).

The monoclinic phase is present in a much lower fraction when using the hydrothermal synthesis route as compared to the calcination route. This is assumed to be an effect of the crystallite size, which influences the stability of the crystal phases, as described above. As a result, the tetragonal phase shows the largest fraction at the lowest doping fraction used here, from where it decreases in a zig-zag shape. The cubic phase increases from about 20% at 2 mol% Eu^{3+} to about 85% at a doping level of 16 mol% but decreases slightly afterwards. Multiple parameters influence the phase fraction behavior here. Firstly, the generally hindered phase quantification in a mixed t-/c-ZrO_2 system, which has been discussed in chapter 3.3.2.1 in more detail, is even more challenging here, due to the small crystallite sizes obtained from the low temperature synthesis routes causing additional peak broadening of the diffraction pattern. Secondly, a slight change in the crystallite growth ratio due to small changes in the system, such as the exact pH during hydrothermal synthesis, can affect the resulting crystal phase composition. The overall behavior, margins of error considered, seems to be comparable to the results obtained from the calcination method, however. To confirm the similarities to the calcined samples more in-depth studies would be necessary, including spectroscopic investigations to assure a comparable dopant and host environment.

3.2.5. Conclusions of the Trivalent Cation Incorporation Studies into Zirconia

The incorporation capacities of trivalent REE could be shown to depend on the crystal phase of zirconia. While the stabilized phases, i.e. t- and c-ZrO_2 are highly capable of incorporating large amounts of dopants, secondary phase formation can be observed when adding a dopant to m-ZrO_2. Nano-sized clusters of the dopant's oxide are assumed to form inside the zirconia matrix. In a HLW repository, where incorporation into zirconia could take place via a dissolution/ recrystallization mechanism different processes could take place. The formation of the dopant oxide is caused by the high calcination temperature in this synthesis approach. However, in a repository, where much lower temperatures will be present, actinides which are not incorporated might not be present as an oxide but would rather be adsorbed to the surface of zirconia. However, the results of the low temperature synthesis studies, which were performed to enable better predictions about the processes potentially occurring in a HLW repository, show the formation of very small crystallites which increase with synthesis time. As could be shown here and was known before, these small crystallites are stable in the high temperature polymorphs, i.e. either t- or c-ZrO_2 rather than the monoclinic phase. This could facilitate the incorporation of actinides into the corrosion layer of Zircaloy cladding which would then hinder the phase transition to the monoclinic phase, once the crystallites grow larger.

The incorporation of REE into stabilized zirconia yielded three different dopant sites, i.e. two fully incorporated sites with 8-fold or 7-fold oxygen coordination and one surface associated site. With increasing doping, the crystallites were shown to decrease in size, likely due to stress induced by the largely oversized dopants. Therefore, the crystallite size will depend on two parameters, which is the crystallite growth rate and the amount of dopant.

The crystallite size, as well as the location of an incorporated actinide, whether it is fully incorporated into the bulk or at a surface layer, will influence the long-term stability of a actinide-zirconia solid solution formed in a HLW repository, and therefore, the contribution of the zirconia corrosion layer to the long-term safety of a HLW repository.

4. Conclusions

In this work, the sorption and incorporation behavior of zirconia towards trivalent metal ions was investigated with a systematic approach using different macroscopic, spectroscopic and diffraction methods, i.e. SEM, TEM, TRLFS, EXAFS, and PXRD. The results reveal a large retention potential of zirconia. The sorption of the trivalent lanthanide Eu^{3+} starts already in the acidic pH range ($\sim$ 3.5), and a complete sorption of Eu^{3+} was achieved from a pH of 5.5, given that the surface area is large enough. Due to the similar electronic configurations and cation radii of Am^{3+} and Pu^{3+}, it can be assumed that these trivalent actinides could behave similar and could therefore be retained on a zirconia corrosion layer on the Zircaloy cladding. The speciation of Cm^{3+} on zirconia could be derived for the first time in Eibl et al.[126] with a combination of batch sorption and TRLFS measurements, and surface complexation modeling was performed to support the derived speciation. Speciation data and modeling data are crucial for being able to predict the long-term safety of a final repository. Therefore, the data on the zirconia system derived here, will be of importance in the safety analysis of a HLW repository.

The incorporation of large fractions of highly oversized cations, such as REE or actinides, into zirconia was known to be possible for a long time. However, the amount of systematic studies of the host and dopant behavior over a large doping range is rather limited. Furthermore, these studies often solely base on EXAFS as spectroscopic method and tend to focus on the stabilized phases only.

The results obtained from studying the behavior of doped $m\text{-}ZrO_2$ is of high importance due to the limited attention of former studies on the monoclinic phase. The monoclinic phase could be, however, of importance to understand the incorporation of actinides in a HLW repository. Here, a slow process of incorporation over a long time period can be assumed as opposed to the direct incorporation of large quantities of dopant at once. The low amount of dopant incorporation would result in a prevailing monoclinic phase. The application of Eu^{3+} and Cm^{3+} TRLFS was performed to clarify the formation processes at low doping levels. It could be seen that even at the lowest doping fraction used here (25 ppm), two distinct phases are present. A share of the dopant could be incorporated into zirconia, which is assumed to be of the non-stabilized monoclinic phase. However, the remaining dopant fraction formed a phase with a much more defined environment, exhibiting comparable properties to the oxide of the respective cation (Eu_2O_3 or Cm_2O_3 simulated by doping Gd_2O_3 with Cm^{3+}). Therefore, it was assumed that nanoclusters of oxides in the zirconia matrix are forming.

TRLFS studies showed, that also in stabilized single-phase systems multiple species are present simultaneously i.e. two different incorporation species, differing in their oxygen coordination sphere as well as a surface associated species in stabilized zirconia. With techniques less specialized on distinguishing multiple species present in one system, such as EXAFS, these species were not detected and no change of the respective signal could be seen when the ratios of the individual species were varied in the doping row. To enable predictions about the stability and the behavior of such incorporated species, the exact speciation is of importance, as a surface associated species might remobilize more easily than a bulk incorporated species. Additionally, the location of oxygen vacancies has implications in the field of material science, where zirconia is a promising material for example as solid electrolyte for solid oxide fuel cells due to its oxygen conductivity and electron insulator properties. The location of oxygen vacancies relative to the dopant and the host is assumed to be crucial for the oxygen mobility in these materials.

For the immobilization of waste streams from SNF reprocessing, zirconia shows very promising properties, as a large quantity of oversized trivalent cations can be hosted in the zirconia matrix due to the formation of oxygen vacancies in the structure.

The capabilities of the zirconia corrosion layer to act as an additional retention barrier are large, although often overlooked in the safety analysis of a potential final repository. This book contains many reasons why this material needs to be considered in this analysis and delivers the basis for being able to do so. At the same time the large amount of research spent on zirconia as a material as well as the research performed here shows the high complexity of this system and the potential for additional detailed studies in some areas. This includes the behavior in hydrothermal synthesis conditions, the stability of actinide doped zirconia under irradiation present due to the proximity to the spent fuel rods, as well as clarification towards the potential of incorporation into degraded surface layers.

5. Experimental Details

5.1. Sorption and Speciation Studies of Eu^{3+} and Cm^{3+} on Zirconia

5.1.1. Material Pre-treatment

The m-ZrO_2 used for the surface sorption studies was obtained from ChemPUR. The pristine material was found to contain a significant amount of organic and inorganic carbon impurities as determined by Total Carbon (TC) measurements. Therefore, the material was purified by calcination for 5 h at 1000° C under ambient atmosphere in a ceramic crucible. A heating rate of 200 °C/h was used. After the completed calcination time, the oven was switched off and left to cool. The calcined product was ground using an agate mortar and pestle.

5.1.2. Bulk Characterization Methods

PXRD

The phase purity and crystallinity of the heat-treated m-ZrO_2 was determined by powder X-ray diffraction (PXRD) using a MiniFlex 600 (Rigaku) diffractometer. About 10 mg of the sample was placed on a low background silicon sample holder and the 2θ range from 5 ° to 90 ° was scanned with a step size of 0.02 ° and a scan speed of 2 °/min.

To obtain information on the phase composition the Rietveld method was used with the PDXL software (version 2) from Rigaku.

For the determination of the reflex FWHMs the average FWHM of the two most intense reflexes at 28.2 ° $(11\bar{1})_m$ and 31.5 ° $(111)_m$ were used.

BET

The specific surface area of the zirconia powder was investigated using N_2-BET physisorption method. These studies were done using a Micromeritics Gemini V, model 2365.

SEM

The morphology and particle size as of the zirconia particles was investigated with SEM. For the SEM investigations, a Quanta 650 SEM of the company ThermoFisher Scientific was used.

Zeta-potential measurements

To investigate the surface properties of the ZrO_2 sample, zeta-potential measurements were conducted with a Zetasizer Nano (Malvern Panalytical). Due to the surface basicity of ZrO_2, carbon dioxide has a tendency to adsorb on the solid, which alters the surface charge of the mineral. Thus, zeta-potential measurements of the solid were conducted in the absence and presence of atmospheric CO_2. The sample rows were prepared using a mineral concentration 0.5 g/L ZrO_2 in 10 mM $NaClO_4$ as background electrolyte in the pH range of 3 - 11. All measurements were done after an equilibration time of three days.

5.1.3. Eu^{3+} Batch-sorption Experiments on ZrO_2

Batch sorption experiments, examining the sorption of Eu^{3+} on m-ZrO_2 as a function of pH were performed. All samples and reagents were prepared in a glove box under nitrogen atmosphere (O_2 < 1 ppm) to exclude any influence from atmospheric CO_2. Carbonate-free Milli-Q water was used in the preparation of all solutions. 10 mM $NaClO_4$ was used as background electrolyte. Several different Eu^{3+} concentrations in the range from $1{\cdot}10^{-7}$ M to $1{\cdot}10^{-5}$ M were used and the solid concentrations were varied from 0.5 g/L – 5.0 g/L. A summary of the experimental conditions used in these investigations is given in Table 10.

Table 10: Summary of experimental conditions used in the batch sorption investigations. 10 mM $NaClO_4$ was used as background electrolyte in all experiments.

Eu^{3+} concentration / M	Solid concentration / g/L	Metal to solid ratio / mol/g
$1{\cdot}10^{-7}$	0.5	$2{\cdot}10^{-7}$
$6{\cdot}10^{-6}$	2.0	$3{\cdot}10^{-6}$
$1{\cdot}10^{-5}$	0.5	$2{\cdot}10^{-5}$

The samples were shaken for a minimum period of three days. Samples were separated by centrifugation with a relative centrifugal force of 4025×g for one hour. The Eu^{3+} concentration in the supernatant was measured with inductively coupled plasma-mass spectrometry (ICP-MS).

For the ICP-MS analysis, the samples were measurements either on an iCAP RQ ICP-MS (Thermo Scientific) or a NexION 350 ICP-MS (Perkin Elmer). The error bars on the sorption edges are based on the standard deviation from triplicate samples.

5.1.4. TRLFS Studies of the Cm^{3+} Surface Complexation on ZrO_2

The spectroscopic investigations of the Cm^{3+} surface complexation and *in situ* speciation were done at a Cm^{3+} concentration of $5\cdot10^{-7}$ M and a solid concentration of 0.5 g/L in all samples if not specified differently.

The pH dependent sorption and speciation of Cm^{3+} on ZrO_2 was investigated with parallel samples containing Cm^{3+}, ZrO_2, and 10 mM $NaClO_4$ as electrolyte. The pH of the samples was increased stepwise from pH 3 to pH 12. After each pH adjustment the samples were equilibrated for two days under constant stirring. For the TRLFS measurements the samples were pipetted into cuvettes, which were sealed in the glove box. After the measurements, the samples were brought back into the glove box and pipetted back to the sample vials, followed by pH adjustments of the complete sample volume.

The TRLFS sorption analyses were performed using a pulsed dye-laser (NarrowScan, Radiant Dyes) with a 1:1 dye mixture of Exalite 389 and Exalite 398, coupled to a Nd:YAG (Continuum, Surelite) pump laser. The Cm^{3+} luminescence emission spectra were recorded between 570 – 640 nm, 1 μs after the exciting laser pulse with a wavelength of 396.6 nm. The laser pulse energy was measured by a pyroelectric energy sensor and was found to be between 2 – 3 mJ in each measurement. Luminescence emission was detected by an optical multichannel analyzer (Shamrock 303i) with 300, 600, or 1200 lines/mm grating and an ICCD-Camera (iStar, Andor). Lifetime measurements were performed by monitoring the luminescence emissions with 5 – 20 μs time delay steps between the laser pulse and the camera gating.

5.1.5. Surface Complexation Modeling

To create a model based on a chemically realistic surface speciation, the calculated species distribution on the m-ZrO_2 was used to fit the formation constants ($\log K°$) for the three different surface species, obtained from TRLFS (see section 3.3.1.3 and Figure 16). The number of water molecules in the coordination sphere of the proposed species has been calculated from the recorded luminescence lifetime data using the *Kimura* equation (Eq. 3, see section 2.2.4.2.1). The deprotonation of the surface hydroxyl-entities upon Cm^{3+} sorption on the ZrO_2 surface was tested for various degrees of deprotonation (dissociation of only one hydroxyl-entity upon full dissociation of all participating surface OH-groups). The reactions described below showed the best fits and were therefore used in the final calculations.

$$\text{Species 1:} \quad 2\ ZrOH + Cm^{3+} + 7\ H_2O \rightarrow (ZrO)_2Cm(H_2O)_7^+ + 2\ H^+$$
$$\text{Species 2:} \quad 3\ ZrOH + Cm^{3+} + 6\ H_2O \rightarrow (ZrO)_3Cm(H_2O)_6^{\pm 0} + 3\ H^+$$
$$\text{Species 3:} \quad 5\ ZrOH + Cm^{3+} + 4\ H_2O \rightarrow (ZrO)_5Cm(H_2O)_4^{2-} + 5\ H^+$$

The surface site density (SSD) of ZrO_2 was calculated using a crystallographic approach. The morphology of ZrO_2 crystals was calculated with the BFDH (Bravais-Friedel-Donnay-Harker) model using the software package 'Mercury CSD 3.9' [179] and the crystallographic data from Smith and Newkirk [180]. For each surface plane, the number of oxygen atoms per square-nanometer was determined assuming that each surface oxygen atom is equivalent to one hydroxylic group that is capable for metal-ion binding reactions. Only the oxygen atoms of the Zr-O-octahedron corners in the uppermost layer were considered, since these are directly accessible for a sorption reaction. For the overall SSD value, the mean SSD value of all planes was calculated and weighted by the relative BFDH area of each plane. For the surface protolysis reactions, the data from Blackwell and Carr [138] was used. For the fitting procedure the parameter estimation software 'UCODE2014' [181] was coupled with the geochemical speciation code 'PHREEQC' [131] using the aqueous speciation and mineral solubility data based on the 'PSI/Nagra Chemical Thermodynamic Database 12/7' [182].

5.2. Incorporation Studies of Eu^{3+}, Y^{3+}, and Cm^{3+} into Zirconia

5.2.1. Materials

Various materials have been used in these studies which are listed below (Table 11).

For Cm^{3+} containing samples, a Cm^{3+} stock solution in HCl, containing 97.2% ^{248}Cm, 2.8%, ^{246}Cm, and less than 0.01% ^{244}Cm was used.

For the experiments, using Th^{4+} doping, Th_{nat} stock solution was prepared via dissolution of $Th(NO_3)_4$ in HCl_{conc} followed by multiple evaporation steps of most of the liquid and re-addition of HCl_{conc} to remove any nitrates from the thorium source.

Table 11: Overview of reagents used in these studies.

Reagent	Manufacturer	Purity
HCl_{conc}	CarlRoth	ROTIPURAN p.a.
NaOH pellets	Sigma-Aldrich	$\geq 98\%$
NaCl	CarlRoth	$> 99\%$ p.a.
$ZrOCl_2 \cdot 8\ H_2O$	Sigma-Aldrich	$> 99.5\ \%$
$EuCl_3 \cdot 6\ H_2O$	Sigma-Aldrich	99.9% purr.
$Lu(NO_3)_3 \cdot xH_2O$	Alfa Aeser	99.99%
Gd_2O_3	Sigma-Aldrich	99.99%, metal basis
$YCl_3 \cdot 6\ H_2O$	Sigma-Aldrich	99.99%
α-Cellulose powder	Sigma-Aldrich	-

5.2.2. Synthesis Methods

5.2.2.1. Calcination Route

Various doped zirconia phases of several sample compositions were synthesized in these studies for various analytic methods using the following procedure.

In the doped ZrO_2 synthesis, 400 mg of $ZrOCl_2 \cdot 8\ H_2O$ was dissolved in 1 mL of 0.01 M HCl. To reach the intended doping level, the acidic Zr^{4+}-containing solution was mixed with an adequate amount of a M^{3+} solution. The acidic metal ion solution (Zr^{4+}, $Ln^{3+}/Y^{3+}/Cm^{3+}/Th^{4+}$) was added dropwise to an alkaline (pH ~ 12) solution of NaCl with a concentration of 0.5 M. The pH was constantly monitored during addition of the acidic precursor solution and readjusted to 12 with NaOH solution when necessary. After the addition of the whole precursor solution, the suspension was kept at 80 °C for 20 h to allow for complete precipitation. Thereafter, the suspension was separated by means of centrifugation and washed twice with 25 mL of Milli-Q water. The residual was dried at 80° C for 20 h and then transferred into an alumina crucible for calcination at 1000° C for 5 h with a heating ramp of 16 K/min and a cooling rate of 0.7 K/min. The resulting doped ZrO_2 solid was mortared into a fine powder.

All samples are named after the main crystal phase (i.e. m-, t- or c-ZrO_2) and the doping fraction. In the case of Y^{3+}/Eu^{3+} co-doped samples, the given doping fraction stands for the overall doping fraction, where 0.5 mol% Eu^{3+} is added in any case.

The additional oxide samples, i.e. Eu_2O_3 and Cm^{3+} doped Gd_2O_3 were prepared via the same (co)-precipitation method described above, only that no Zr-precursor was used. All further treatments are equivalent to the preparation of the doped zirconia phases.

5.2.2.2. Low Temperature Route

The low temperature route follows the same (co)-precipitation method as described above. However, after adding the whole metal precursor solution to the alkaline electrolyte solution, the resulting suspension was heated at temperatures between 80 and 200 °C for 8 h – 11 d. The heat treatment at temperatures below 100 °C were conducted in sealed Greiner tubes, while hydrothermal bombs consisting of a Teflon vessel surrounded by a steel liner were used for higher temperatures. After the desired synthesis time, the suspension was taken out of the oven to cool down. When cool, the suspension was separated by means of centrifugation and the solid was washed twice with 25 mL of Milli-Q water. The resulting powder was dried at 80 °C over night and mortared to receive a fine powder.

5.2.3. Bulk Characterization Methods

PXRD

The phase composition of the synthetic doped zirconia phases was analyzed with PXRD. The samples containing solely Eu^{3+} as well as the ones containing solely Lu^{3+}, Gd^{3+} or Y^{3+} were measured at the Rossendorf beamline (ROBL) at the European synchrotron research facility (ESRF) in Grenoble, France. The measurement setup has a rotating capillary sample holder and a Pilatus 2M 2D detector. The X-ray wavelength was set to 0.73804 Å. Each sample was measured for a total of 10 min.

All other calcined samples were analyzed using a MiniFlex 600 (Rigaku) using the same measurement setup as described in chapter 5.5.1.2.

The samples from low-temperature synthesis were measured using a D8 Advance (Bruker) diffractometer, equipped with a Göbel Mirror a Lynxeye detector. The X-ray tube had a copper source, providing Cu $K_{\alpha 1}$ X-rays. Sample preparation involved grinding ~0.1 g of sample material, mixed with ~1 ml of amyl acetate, in a pestle and mortar. The resultant slurries were transferred to glass microscope slides and air dried. Samples were scanned from 5 - 90° 2θ angle, with a step size of 0.02° and a count time of 0.2 s per step.

For details about the crystallite size determination and FWHM determination the reader is referred to chapter 5.5.1.2. The lattice parameters were derived from *Rietveld* refinement.

For the determination of the doping fractions based on the lattice parameters, the following equation (Eq. 5) of *Freris et al.*[145] was used, which was derived by fitting the cell volume of the tetragonal and cubic phase over the doping fractions. V_{cell} is the unit cell volume of the tetragonal phase, 133.85 Å³ is twice the cell volume of the pristine, undoped tetragonal ZrO_2 and 0.223 Å is the slope of the line fitted by *Freris et al.* in a plot of the cell volume over the doping fraction.

$$\%Eu = \frac{2V_{cell} - 133.85 \text{ Å}^3}{0.223 \text{ Å}} \qquad (5)$$

For the cubic lattice the following equation (Eq. 6) was applied:

$$\%Eu = \frac{V_{cell} - 133.85 \text{ Å}^3}{0.223 \text{ Å}} \qquad (6)$$

The factor two used for the calculation for the tetragonal phases results from the fact that the tetragonal unit cell contains half the amount of cations than the cubic unit cell. By multiplying with two the fit of the cell volume over the doping fraction can be derived in one linear fit.

SEM and TEM

SEM and TEM were combined to investigate both, the particle morphology and size of selected ZrO_2 samples, as well as to confirm a homogeneous distribution of the dopant in the material. The SEM images were obtained on a FEI Quanta 650 with samples either as powder or from suspension on a stub holder. TEM measurements were performed on a FEI Tecnai F30 where the samples were applied from suspension on a TEM grid.

5.2.4. EXAFS Studies of the Local Environment of Y^{3+} in ZrO_2

For the EXAFS studies of the low doping range, yttrium doped zirconia was prepared following the synthesis method described above. The high doping range was studied by preparing Y^{3+}/Eu^{3+} co-doped samples. All samples were diluted with α-cellulose powder to an analyte concentration of 1 mass% and pressed into a pellet with a handheld press. The pellets were placed in a polyethylene confinement for the EXAFS measurements. The measurements were performed at the INE-Beamline[183,184] at the KIT synchrotron source, Karlsruhe, Germany at the Zr-K- and Y-K-edge at room temperature in transmission mode. The storage ring operating conditions were 2.5 GeV and 80 - 120 mA. A Ge [422] double crystal monochromator coupled with collimating and focusing Rh-coated mirrors was used. XAFS spectra at the Zr K- (17.998 keV) and at the Y K-edge (17.038 keV) were recorded in transmission mode using Ar-filled ionization chambers at ambient pressure. For energy calibration, XAFS spectra of Zr and Y metal foils were recorded simultaneously. The data was analyzed using the EXAFSPAK software suite[185]. Phases and amplitudes of the interatomic scattering paths were calculated using FEFF8.20.[186] A spherical cluster of atoms with a radius of 6 Å using the monoclinic structure (space group $P2_1/c^{127}$) or the fluorite-type structure (space group $Fm\bar{3}m^{187}$) was used in the FEFF calculation.

5.2.5. Eu^{3+} TRLFS Incorporation Studies

UV-excitation

TRLFS was utilized to record excitation spectra, emission spectra, and luminescence lifetimes of Eu^{3+} incorporated in zirconia. UV-excitation (λ_{ex} = 394 nm) experiments at room temperature of all synthetic Eu^{3+} or Eu^{3+}/Y^{3+} co-doped samples were performed to gain an insight into the overall Eu^{3+} symmetry in the solid matrices by simultaneous excitation of all present Eu^{3+} species in the samples. In these experiments, the ground Eu^{3+} doped ZrO_2 powders were measured in small polypropylene vials. These experiments were conducted with a Nd:YAG pump laser coupled to an OPO system, which was tuned to an output wavelength of 394 nm. The pulse energy was measured by a pyroelectric energy sensor and was found to be between 2 – 3 mJ.

Eu^{3+} luminescence emission spectra were recorded between 570 - 650 nm, 1 µs after the exciting laser pulse. Luminescence emission was detected by an optical multichannel analyzer (Shamrock 303i) with 300, 600 or 1200 lines/mm grating and an ICCD-Camera (iStar, Andor). Lifetime measurements were performed by monitoring the luminescence emissions with 5 µs – 500 µs time delay steps between the laser pulse and the camera gating.

Site-selective excitation

Site-selective excitation of the $^5D_0 \leftarrow {}^7F_0$ transition (~573 - 583 nm) at 10 K was used to study non-equivalent Eu^{3+} environments in the samples, and to improve the overall resolution for an unambiguous determination of the Eu^{3+} site symmetry in the various zirconia polymorphs. The experiments were carried out in copper holders sealed with quartz glass windows, to ensure an adequately low temperature at the sample.

For these experiments, a pulsed dye-laser (NarrowScan, Radiant Dyes) with Rhodamine 6G dye, coupled to a Nd:YAG (Continuum, Surelite) pump laser was used. The excitation wavelength was varied between 570 and 585 nm, the pulse energy was between 4 – 5 mJ here. Measurements were conducted at temperatures below 10 K, using a He-refrigerated cryostat (Janis and Sumitomo, SHI cryogenics group). The detection of the luminescence was performed as described above.

5.2.6. Cm^{3+} TRLFS Incorporation Studies

UV-excitation

The Cm^{3+} TRLFS measurements were performed using UV-excitation (λ_{ex} = 396.6 nm) at room temperature. The measurement conditions and setup were identical to the ones described in the first paragraph of chapter 5.5.2.5, however, the output wavelength was tuned to 396.6 nm here.

Site-selective excitation

The site-selective measurements of Cm^{3+} were performed with the same setup as for Eu^{3+}, but with different laser dyes were used. Due to the broad emission and excitation range of these samples, four different dyes, Rhodamine 6G, Sulforhodamine, Rhodamine 101, and DCM, had to be used to cover the whole range. The excitation wavelength was varied between 580 nm and 660 nm.

Appendix

EXAFS Fitting Results using Various Fitting Approaches

Table A1 presents the resulting CNs, where the abs–O path was fitted.

Table A1: Structural parameters derived from the k^3–weighted EXAFS spectra for the Zr and the Y K–edge. Fixed CN are marked with an asterisk (*). Abs–O CN is fitted here. Fitting range is $2 - 10.5$ Å^{-1}, R: radial distance, error ± 0.01 Å, CN: coordination number, error $\pm 20\%$, σ^2: Debye–Waller factor, error ± 0.0005 Å, amplitude reduction factor (S_0^2) = 1.0, number of independent points = 16.

Y^{3+} content / mol%		All shells fitted, CN of M–O fitted, other CNs fixed							
		Zr K–edge			R–value / %	Y K–edge			R–value / %
	Path	R / Å	CN	σ^2 / Å^2		R / Å	CN	σ^2 / Å^2	
16	abs–O	2.14	6.13	0.009	8.30	2.32	7.9	0.010	6.94
	abs–M	3.54	12*	0.014		3.61	12*	0.009	
	abs–O–M	3.90	48*	0.020		4.20	48*	0.008	
	abs–O	4.52	24*	0.101		4.50	24*	0.033	
18	abs–O	2.15	7.70	0.012	7.06	2.31	7.0	0.009	5.72
	abs–M	3.56	12*	0.014		3.61	12*	0.010	
	abs–O–M	3.93	48*	0.025		4.20	48*	0.005	
	abs–O	4.36	24*	0.044		4.48	24*	0.036	
24	abs–O	2.16	7.03	0.011	7.68	2.32	8.2	0.013	6.14
	abs–M	3.56	12*	0.013		3.61	12*	0.010	
	abs–O–M	3.83	48*	0.028		4.20	48*	0.006	
	abs–O	4.41	24*	0.055		4.49	24*	0.030	
26	abs–O	2.14	7.44	0.011	6.10	2.31	7.2	0.010	5.81
	abs–M	3.55	12*	0.013		3.61	12*	0.010	
	abs–O–M	3.93	48*	0.028		4.10	48*	0.012	
	abs–O	4.35	24*	0.050		4.47	24*	0.043	

The fitting results, where the CN of the abs–M path is fitted, are presented in Table A2.

Table A2: Structural parameters derived from the k3–weighted EXAFS spectra for the Zr and the Y K–edge. Fixed CN are marked with an asterisk (*). Abs–M CN is fitted here. Fitting range is 2 – 10.5 Å–1, R: radial distance, error ± 0.01 Å, CN: coordination number, error ± 20%, σ2: Debye–Waller factor, error ± 0.0005 Å, amplitude reduction factor (S02) = 1.0, number of independent points = 16.

Y^{3+} content / mol%		All shells fitted, CN of M–M fitted, other CNs fixed							
		Zr K–edge			R–value / %	Y K–edge			R–value / %
	Path	R / Å	CN	σ^2 / Å^2		R / Å	CN	σ^2 / Å^2	
16	abs–O	2.13	8*	0.012	9.65	2.32	8*	0.010	6.67
	abs–M	3.56	13.27	0.014		3.62	18.9	0.011	
	abs–O–M	3.93	48*	0.006		4.01	48*	0.006	
	abs–O	4.37	*24*	0.086		4.59	24*	0.064	
18	abs–O	2.15	8*	0.013	7.42	2.31	8*	0.011	5.94
	abs–M	3.56	9.97	0.013		3.62	18.36	0.012	
	abs–O–M	3.61	48*	0.033		4.04	48*	0.009	
	abs–O	4.49	24*	0.068		4.60	24*	0.060	
24	abs–O	2.15	8*	0.013	8.82	2.32	8*	0.012	6.19
	abs–M	3.56	9.30	0.012		3.61	11.99	0.010	
	abs–O–M	3.66	48*	0.033		4.17	48*	0.006	
	abs–O	4.47	24*	0.064		4.49	24*	0.030	
26	abs–O	2.14	8*	0.011	6.63	2.31	8*	0.012	6.59
	abs–M	3.55	10.97	0.013		3.61	18.52	0.012	
	abs–O–M	3.58	48*	0.038		4.01	48*	0.004	
	abs–O	4.52	24*	0.073		4.60	24*	0.083	

In Table A3 the CN of the abs–O and the abs–M path is fitted.

Table A3: Structural parameters derived from the k^3–weighted EXAFS spectra for the Zr and the Y K–edge. Fixed CN are marked with an asterisk (*). Abs–O and Abs–M CN is fitted here. Fitting range is 2 – 10.5 Å^{-1}, R: radial distance, error ± 0.01 Å, CN: coordination number, error ± 20%, σ^2: Debye–Waller factor, error ± 0.0005 Å, amplitude reduction factor (S_0^2) = 1.0, number of independent points = 16.

Y^{3+} content / mol%	Path	Zr K–edge R / Å	CN	σ^2 / Å^2	R–value / %	Y K–edge R / Å	CN	σ^2 / Å^2	R–value / %
16	abs–O	2.14	5.75	0.008		2.32	8.0	0.010	
	abs–M	3.57	13.85	0.013	7.64	3.62	19.26	0.011	6.60
	abs–O–M	3.94	48*	0.002		4.01	48*	0.006	
	abs–O	4.57	24*	0.104		4.6**	24*	0.062	
18	abs–O	2.15	7.60	0.012		2.31	7.06	0.009	
	abs–M	3.56	9.88	0.013	6.94	3.62	17.83	0.012	5.77
	abs–O–M	3.63	48*	0.032		4.05	48*	0.008	
	abs–O	4.49	24*	0.070		4.6**	24*	0.06**	
24	abs–O	2.16	7.41	0.011		2.32	8.16	0.013	
	abs–M	3.55	10.91	0.013	7.33	3.61	12.06	0.010	6.14
	abs–O–M	3.68	48*	0.028		4.17	48*	0.064	
	abs–O	4.48	24*	0.067		4.49	24*	0.030	
26	abs–O	2.14	7.41	0.011		2.31	6.93	0.010	
	abs–M	3.55	10.91	0.013	5.86	3.62	16.97	0.011	6.36
	abs–O–M	3.61	48*	0.039		4.01	48*	0.002	
	abs–O	4.52	24*	0.078		4.94	24*	0.026	

All shells fitted, CN of M–O and M–M fitted, other CNs fixed

Comparison of EXAFS Fitting Results to Crystallographic Data

In Table A4, a comparison between the crystallographic data of m-ZrO_2 and c-ZrO_2 to the average EXAFS parameters derived from fitting is presented. While the Zr-O distance in m- and c-ZrO_2 as well as the Zr-Zr distance in c-ZrO_2 are comparable between the different data sets the Zr-Zr distance in m-ZrO_2 is significantly shorter in EXAFS. This is likely caused by the aforementioned problem of fitting this monoclinic structure which, as the crystallographic parameters reveal, has a variety of individual atomic distances and, therefore, multiple individual scattering paths. Since fitting every single one would not yield a reasonable fit, the Zr-Zr shell was fitted using three sub-shells. This compromise, while necessary, will, however, reduce the precision of the fitting result to some extent. Furthermore, the multitude of scatter paths also causes a variety of multiple scatter paths overlapping the single scatter paths and therefore, hindering the fit to the structure model.

Table A4: Comparison of crystallographic parameters of undoped m-ZrO$_2$[127] and c-ZrO$_2$[187] with the average parameters derived by EXAFS over the fitting range of 0.5 - 2 mol% (m-ZrO$_2$) and of 16-26 mol% (c-ZrO$_2$).

	Crystallographic data	
Bond	Zr-O	Zr-Zr
m-ZrO$_2$	2.0511, 2.0702, 2.1358, 2.1466, 2.1681, 2.2555, 2.2715	3.3441, 3.442 x 2, 3.4568 x 2, 3.4745, 3.5783, 3.9262 x2, 4.0238 x 2, 4.5414
Average	2.1570 x 7	3.7197 x 12
c-ZrO$_2$	2.224 x 8	3.631 x 12
	Average EXAFS fitting results	
m-ZrO$_2$	2.14 x 7	3.47 x 7, 3.67 x 4, 3.95 x 1
Average	2.14 x 7	3.58 x 12
c-ZrO$_2$	2.15 x 8	3.55 x 12

Calculation of Number of Oxygen Vacancies

The number of oxygen vacancies in dependence of the trivalent doping fraction is presented in Figure A1 along with the amount of Zr and Eu atoms with an oxygen vacancy in the first coordination sphere when assuming a statistical distribution with no preferential oxygen vacancy location. Since the oxygen atoms in ZrO$_2$ are coordinated to four cations, every oxygen vacancy will influence four Zr and/or dopant ions in the crystal lattice. It further implies that only up to 50 mol% trivalent dopant can be theoretically hosted without more than one oxygen vacancy in each cation NN coordination. Therefore, values at higher doping levels are larger than 100, indicating that the presence of two oxygen vacancies in some cation NN coordination spheres are necessary.

The number of Eu^{3+} cations per 100 overall cations (Eu+Zr(+Y)) with one oxygen vacancy (O$_{vac}$), assuming statistical follows the equation:

$$N\left(Eu_{O_{vac}}\right) = 200 * x(Eu)^2 \qquad (Eq.\,S1)$$

The number of Zr^{4+} cations per 100 overall cations (Eu + Zr (+ Y)) with one O$_{vac}$, assuming statistical distribution follows the equation:

$$N\left(Eu_{O_{vac}}\right) = 200 * (x(Eu) - x(Eu)^2) \qquad (Eq.\,S2)$$

The number of O$_{vac}$ follows equation S3.

$$N(O_{vac}) = 50 * x(Eu) \qquad (Eq.\,S3)$$

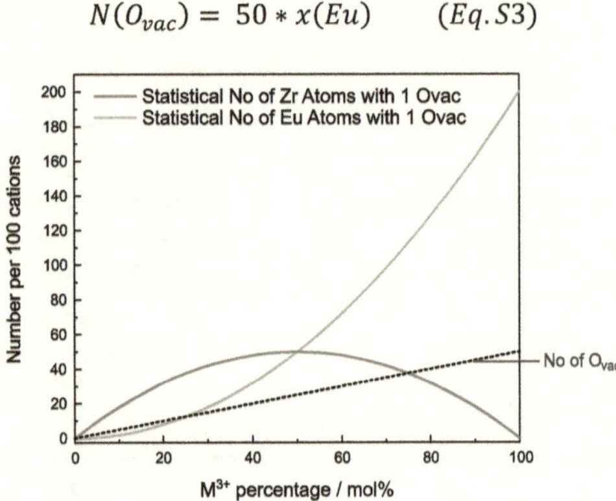

Figure A1: Number of oxygen vacancies created in the ZrO₂ lattice in dependence on the M³⁺ doping fraction (black, dotted line) and number of cations with an oxygen vacancy in the first coordination sphere of Eu (orange line) or Zr (red line), assuming a non-preferential, statistical distribution. Numbers larger than 100 result from the necessity of two oxygen vacancies in the coordination sphere of one cation.

Defect Luminescence in Doped ZrO₂

In many of the emission spectra of the Eu³⁺ doped zirconia samples, especially at low doping percentages, a peculiarity can be seen in the emission spectra, as marked in Figure A2, left.

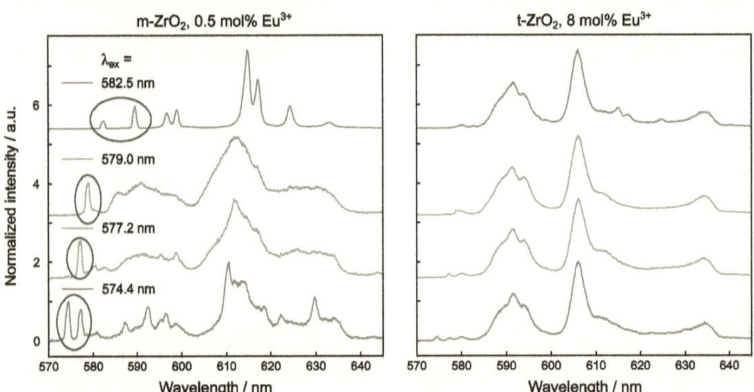

Figure A2: Emission spectra of a monoclinic zirconia sample doped with 0.5 mol% Eu³⁺ which shows unusual peaks in the lower wavelength region (marked with circles, left) and t-ZrO₂ doped with 8 mol%, where no such behavior is visible in comparison (right).

These peaks appear as very sharp peaks in the high energy region, where, at some excitation wavelengths, two peaks can be seen while in others only one such peak is visible. Furthermore, the peak width of these sharp peaks is in contrast to the very broad emission peaks observed in the remaining parts of the spectra, especially when excitation is performed in the broad excitation peaks 2 and 3. When the Eu³⁺ fraction is increased no such peaks can be observed as presented in Figure A2, right for a sample doped with 8 mol% Eu^{3+}. A non Eu^{3+} doped sample does not show any comparable peaks.

However, when using a small fraction of Eu^{3+} doping (0.5 mol%) with a large fraction of a co-dopant, such as Y^{3+} (17.5 mol%), this effect becomes very dominant over a large range of excitation wavelengths (Figure A3, left). The peak positions and intensities are dependent on the excitation energy as further elucidated in Figure A3, right. These observations lead to the assumption, that a differing luminescence mechanism is responsible for the observed peaks than just the luminescence of Eu^{3+}. Since the intensity of these peaks, however, correlates with the doping fraction, it is likely that the effect correlates with the amount of defects in the lattice. Besides the formation of oxygen vacancies, defect electrons can form in defect systems. The presence of defect electrons in ZrO_2 was observed before by Liu et al.[188]. The energetically degenerate electrons create an energy band, which can absorb and emit energy.

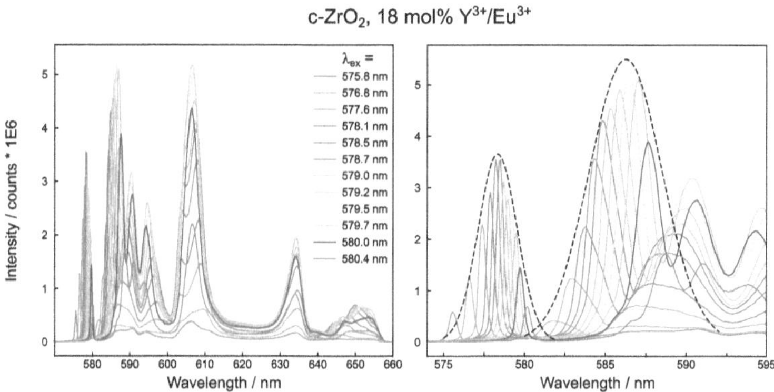

Figure A3: Visualization of the defect site luminescence of an example sample of a Y^{3+} and Eu^{3+} co-doped ZrO_2 sample for various excitation energies plotted for the range of the $^7F_0$ transition to the $^7F_3$ transition (left) and in a reduced area where only $^7F_0$ and $^7F_1$ emission can be seen, including dashed lines to visualize the behavior of this emission effect (right).

Since the presence of Eu^{3+} as laser light absorbing cations seems to be necessary to observe these defect electron effect, an energy transfer mechanism of absorbing Eu^{3+} to the defect electrons, followed by emission of luminescence by the defects or back transition to Eu^{3+} sites, is assumed.

As a result of this effect, the determination of the splitting of the J levels is hindered. However, since it is expected that the excitation of the defect electrons is caused by the emission of the Eu^{3+}, the sharp peak of the defect luminescence emission covers up the underlying original emission of Eu^{3+}. Therefore, these peaks caused by the defect electrons are, nevertheless, a result of Eu^{3+} luminescence at this very wavelength and hence, part of the Eu^{3+} splitting pattern.

www.ingramcontent.com/pod-product-compliance
Lightning Source LLC
LaVergne TN
LVHW041727190726
843493LV00007B/2250